AF829781

Palgrave Studies in Animals and Literature

Series Editors
Susan McHugh
Department of English
University of New England
Auburn, USA

Robert McKay
School of English
University of Sheffield
Sheffield, UK

John Miller
School of English
University of Sheffield
Sheffield, UK

Various academic disciplines can now be found in the process of executing an 'animal turn', questioning the ethical and philosophical grounds of human exceptionalism by taking seriously the nonhuman animal presences that haunt the margins of history, anthropology, philosophy, sociology and literary studies. Such work is characterised by a series of broad, cross-disciplinary questions. How might we rethink and problematise the separation of the human from other animals? What are the ethical and political stakes of our relationships with other species? How might we locate and understand the agency of animals in human cultures?

This series publishes work that looks, specifically, at the implications of the 'animal turn' for the field of English Studies. Language is often thought of as the key marker of humanity's difference from other species; animals may have codes, calls or songs, but humans have a mode of communication of a wholly other order. The primary motivation is to muddy this assumption and to animalise the canons of English Literature by rethinking representations of animals and interspecies encounter. Whereas animals are conventionally read as objects of fable, allegory or metaphor (and as signs of specifically human concerns), this series significantly extends the new insights of interdisciplinary animal studies by tracing the engagement of such figuration with the material lives of animals. It examines textual cultures as variously embodying a debt to or an intimacy with animals and advances understanding of how the aesthetic engagements of literary arts have always done more than simply illustrate natural history. We publish studies of the representation of animals in literary texts from the Middle Ages to the present and with reference to the discipline's key thematic concerns, genres and critical methods. The series focuses on literary prose and poetry, while also accommodating related discussion of the full range of materials and texts and contexts (from theatre and film to fine art, journalism, the law, popular writing and other cultural ephemera) with which English studies now engages.

Series Board:
Karl Steel (Brooklyn College)
Erica Fudge (Strathclyde)
Kevin Hutchings (UNBC)
Philip Armstrong (Canterbury)
Carrie Rohman (Lafayette)
Wendy Woodward (Western Cape)

Vera Fibisan • Rachel Murray
Editors

Blue Extinction in Literature, Art, and Culture

palgrave
macmillan

Editors
Vera Fibisan
University of Sheffield
Sheffield, UK

Rachel Murray
University of Bristol
Bristol, UK

ISSN 2634-6338 ISSN 2634-6346 (electronic)
Palgrave Studies in Animals and Literature
ISBN 978-3-031-69909-2 ISBN 978-3-031-69910-8 (eBook)
https://doi.org/10.1007/978-3-031-69910-8

© The Editor(s) (if applicable) and The Author(s), under exclusive license to Springer Nature Switzerland AG 2024

This work is subject to copyright. All rights are solely and exclusively licensed by the Publisher, whether the whole or part of the material is concerned, specifically the rights of translation, reprinting, reuse of illustrations, recitation, broadcasting, reproduction on microfilms or in any other physical way, and transmission or information storage and retrieval, electronic adaptation, computer software, or by similar or dissimilar methodology now known or hereafter developed.
The use of general descriptive names, registered names, trademarks, service marks, etc. in this publication does not imply, even in the absence of a specific statement, that such names are exempt from the relevant protective laws and regulations and therefore free for general use.
The publisher, the authors and the editors are safe to assume that the advice and information in this book are believed to be true and accurate at the date of publication. Neither the publisher nor the authors or the editors give a warranty, expressed or implied, with respect to the material contained herein or for any errors or omissions that may have been made. The publisher remains neutral with regard to jurisdictional claims in published maps and institutional affiliations.

This Palgrave Macmillan imprint is published by the registered company Springer Nature Switzerland AG.
The registered company address is: Gewerbestrasse 11, 6330 Cham, Switzerland

If disposing of this product, please recycle the paper.

Acknowledgments

We would like to thank our contributors for their hard work and generous contributions to the collection. It has been inspiring to work with you all. This edited collection started out as an ASLE-UKI seminar in 2021 and we are grateful to all of the invited participants. It then grew into a two-day international conference at the University of Sheffield in July 2022, and we'd like to thank everyone who shared their ideas, whether in-person or online, and contributed to such an enriching few days of discussion.

Our thanks go to the Leverhulme Trust for funding the conference, as well as to the Sheffield Animal Studies Research Centre (ShARC) and the School of English for hosting it. A big thank you to Daniel Bowman for assisting us with the conference.

We'd also like to thank the series editors, John Miller, Bob McKay, and Susan McHugh for their support and encouragement along the way. We are grateful to the production team at Palgrave for their assistance with this volume too.

Vera Fibisan: I am deeply grateful for the support of my colleagues in ASLE-UKI and in the English Department at Sheffield. Also, my co-editor, Rachel Murray, whom I greatly enjoyed working with.

Rachel Murray: I'd like to thank my co-editor Vera Fibisan, who introduced me to the concept of blue extinction back in 2021—it's been wonderful to work together these past few years. I'd also like to thank Katherine Ebury at Sheffield for her support as a mentor while I was based at Sheffield. And Jack and Shrimp, my companions.

Contents

Notes on Contributors

Stacy Alaimo is Professor of English and Core Faculty Member in Environmental Studies at the University of Oregon. Her books include *Undomesticated Ground: Recasting Nature as Feminist Space* (2000); *Bodily Natures: Science, Environment, and the Material Self* (2010), and *Exposed: Environmental Politics and Pleasures in Posthuman Times* (2016). She also co-edited *Material Feminisms* (2008), edited the 28-chapter volume *Matter* (2016) in the Gender series of Macmillan Interdisciplinary Handbooks, and edited a special volume of *SLSA's Configurations on Science Studies and the Blue Humanities.* Her forthcoming book is tentatively titled, *The Abyss at Hand: Aesthetic Encounters with Deep Sea Life.*

Akshita Bhardwaj is a PhD candidate in the Department of Indian and World Literatures at the English and Foreign Languages University, Hyderabad (India). Grounded in Literary Animal Studies, her research attends to nonhuman animal presences in texts—both literary and non-literary—that escape the usual registers of anthropocentric meaning. Her other interest areas are Postcolonial animal studies, Indology studies, and Spatial literary studies.

Rick De Vos conducts research in animal studies and in anthropogenic extinction, in particular the cultural and historical significance of extinction and the way it is articulated and practiced. He is an adjunct research fellow in the Centre for Culture and Technology at Curtin University in Western Australia, and a series editor for Sydney University Press' Animal

Politics series. Rick is a member of the *Extinction Studies Working Group* and has published widely on extinction, and species and colonial histories.

Karen Eckersley researches the intersection between women's surrealism and post-human feminism. Karen has published peer-reviewed articles in the *Journal for Literary and Intermedial Crossings, Transpositiones, Romance Studies,* and *Gothic Nature* where she has published an article on Bishop for their special edition entitled "Haunted Shores." Her most recent work is a book chapter on Leonora Carrington's human-animal hybridity in her writing, published in *Beastly Modernisms: The Figure of the Animal in Literature and Culture* (EUP, 2023).

Vera Fibisan is an Honorary Researcher at The University of Sheffield. Her areas of interest include extinction studies, ecocriticism, ecofeminism, coastal radical landscape poetry, and blue humanities. She is predominantly a practice-based researcher and writer, focusing on bodies of water and marginalia. She has published poetry notably in *The Sheffield Anthology* (Smith/Doorstop, 2012), *CAST: The Poetry Business Book of New Contemporary Poets* (Smith/Doorstop, 2014), the *Wretched Strangers Anthology* (Boiler House Press, 2018), and *Voices for Change Anthology* (2020). She is ASLE-UKI Website and Social Media Officer and Podcast Co-host of *Green Listening: Discussions in Ecocriticism.*

Megan Hayes is a PhD candidate in the University of Oregon's Environmental Sciences, Studies, and Policy program, where she is currently living and working on Kalapuya lands. Her research, writing, and creative practice draw from feminist science studies, the blue humanities, and artistic research methodologies, and are informed by undergraduate studies in Photography and Situated Media at the University of Technology, Sydney, and a Research Master's in Cultural Analysis from the University of Amsterdam. Megan's PhD project, *How to Love an Oyster: Chemistry, Slippage, Attachment*, studies oyster–human relationships and the worlds those relations sustain.

Henry Ivry is a Lecturer in Twentieth and Twenty-First Century Literature in the School of Critical Studies at the University of Glasgow. His research explores how climate and racial justice are two sides of the same coin and how various facets of African American culture and thought form a vital archive for building otherwise worlds. His first book *Transscalar Critique: Climate, Blackness, Crisis* was published by

Edinburgh UP in 2023 and his academic work has appeared or is forthcoming in journals including *Contemporary Literature*, *English Literary History*, *Modern Fiction Studies*, *ISLE: Interdisciplinary Studies in Literature and Environment*, and *English Language Notes*. Henry also works regularly as a freelance music journalist with bylines in *Resident Advisor*, *DJ Mag*, *Bandcamp Daily*, and *Beatportal*.

Max Karpinski is a Social Sciences and Humanities Research Council (SSHRC) Postdoctoral Fellow at York University, Canada. He has guest edited two special issues of *Canadian Literature* on the topic of "Poetics and Extraction," and other recent research can be found in *Canada and Beyond*, *Imaginations: Journal of Cross-Cultural Image Studies*, *ISLE: Interdisciplinary Studies in Literature and Environment*, and *SCL: Studies in Canadian Literature*.

Steve Mentz is Professor of English at St. John's University in New York City. His most recent book, *Sailing without Ahab: Ecopoetic Travels*, appeared in April 2024 from Fordham University Press. He has also written six other books, including *An Introduction to the Blue Humanities* (2023), *Ocean* (2020), and *At the Bottom of Shakespeare's Ocean* (2009). He is also editor or co-editor of seven collections, including *Water and Cognition in Early Modern English Literature* (2024), and *A Cultural History of the Sea in the Early Modern Age* (2021). He blogs at *The Bookfish*, www.stevementz.com and (still) tweets @stevermentz.

Rachel Murray is Lecturer in Literature and the Environment at the University of Bristol. Her research specializes in modernist and twentieth-century literature, animal studies, and the environment. She is the author of *The Modernist Exoskeleton: Insects, War, Literary Form* (EUP, 2020). She is also the co-editor, with Caroline Hovanec, of "Reading Modernism in the Sixth Extinction" (2022) for *Modernism/modernity* Print Plus. Her current book project, *Marine Attachments*, focuses on marine life and ideas of attachment in modern and contemporary poetry.

Dominic O'Key is Leverhulme Early Career Fellow at the University of Sheffield. He is the author of *Creaturely Forms in Contemporary Literature: Narrating the War Against Animals* (Bloomsbury, 2022), which was shortlisted for Association for the Study of Literature and Environment's best academic monograph prize. He is currently working on the cultural politics of wildlife conservation and extinction.

Killian Quigley is a research fellow at the Australian Catholic University's Institute for Humanities and Social Sciences. He was previously the inaugural postdoctoral fellow at the Sydney Environment Institute, University of Sydney. Killian is the author of *Reading Underwater Wreckage: An Encrusting Ocean* (2023) and co-editor, with Margaret Cohen, of *The Aesthetics of the Undersea* (2019). He is cofounder, with Charne Lavery and Laurence Publicover, of the *Critical Seabed Studies* working group, and a member by invitation of the *Oceanic Humanities for the Global South* research network.

Rosanne van der Voet is a lecturer in environmental humanities and urban studies at Leiden University in the Netherlands. She completed her PhD in Creative Writing at the University of Sheffield in 2023 on storytelling and the environmental crisis of the oceans. Her research spans across various interdisciplinary strands of the blue humanities, with particular focus on nonhuman experience of environmental issues, creative-critical approaches, and applied ecocritical analysis of new nature-based water management projects in urban and industrial environments in the Netherlands. Her work can be found in journals such as *ISLE*, *Ecozon@*, *Journal of Posthumanism*, *Book 2.0*, and *The York Journal*.

CHAPTER 1

Introduction: Blue Extinction

Rachel Murray and Vera Fibisan

Set in the near future, Ned Beauman's satirical eco-thriller *Venomous Lumpsucker* (2022) follows a corrupt mining executive, Mark Halyard, and a biologist employed by a deep-sea mining firm, Karin Resaint, as they embark on an unlikely quest to save a creature from extinction. The creature in question is the fictional venomous lumpsucker (*Cyclopterus venenatus*), a "bumpy, greyish fish" with a "toadlike face" (18), an "[u]gly little fucker, with a nasty bite" (23). Having declined in numbers due to anthropogenic pollution, the lumpsucker's last-known breeding grounds in the Baltic Sea are inadvertently destroyed by the mining company early in the text, and its fate is left hanging in the balance. The extinction of this deep-sea fish should trouble no one, for it is a species which, as Halyard states bluntly, "[n]obody cares about" (16). Unlike the last giant panda, Chiu Chiu, whose death twelve years earlier had caused an outpouring of grief "unprecedented in the history of the human race," (13) this unsightly fish is deemed incapable of eliciting the kind of response reserved for so-called

R. Murray (✉)
University of Bristol, Bristol, UK
e-mail: rachel.e.murray@bristol.ac.uk

V. Fibisan
University of Sheffield, Sheffield, UK

© The Author(s), under exclusive license to Springer Nature Switzerland AG 2024

V. Fibisan, R. Murray (eds.), *Blue Extinction in Literature, Art, and Culture*, Palgrave Studies in Animals and Literature,
https://doi.org/10.1007/978-3-031-69910-8_1

charismatic species, whose cute or awe-inspiring nature has often served to galvanise conservation efforts (Lorimer 2007; Heise 2016). Nobody could pretend to miss a creature, thinks Resaint, as "distant from human apprehension as any species that died out with the dinosaurs" (97). The disappearance of the venomous lumpsucker would surely be "[n]o great loss" (23) given that its existence had barely even been registered.

A few things happen, though, to make Halyard and Resaint care deeply about the fate of this obscure fish. The first is that Resaint, who has been hired by the mining company to study the venomous lumpsucker, discovers that the species is highly intelligent, perhaps "the most intelligent fish on the planet" (111). This causes it to gain sudden value as a commodity within the "extinction industry," (14) a free-market system in which companies buy and sell "extinction credits" (22), awarded for supposedly beneficial activities to help save an endangered species. For the unscrupulous Halyard, whose short selling of extinction credits has recently been scuppered by a hack on the biobanks, which hold the data of extinct and endangered species, the creature's survival may avert financial and personal ruin. For Resaint, however, the discovery of nonhuman intelligence that is at once highly sophisticated, yet which has largely gone unobserved, reignites an earlier "epiphany" she'd had while studying a rare parasitic wasp. Having "found it easy not to care about extinction," Resaint is shocked by the indifference of a fellow biologist to news of the wasp's impending demise:

> How could this brilliant, intricate, hilarious thing—the fluke result of an unrepeatable process, the legacy of some dizzying number of past individuals ... not be valuable in itself? What did it matter if anyone appreciated it, if it did any good for anyone? ... A tragedy could only be a tragedy if it hurt somebody? ... The absence, in this case, of the weaker arguments against extinction—wouldn't we be sad to lose this little fellow?—made space for Resaint to feel the truth of this much graver argument. (104)

While the extinction of this parasitic wasp may not hurt anyone, Resaint realises, it would still be a "terrible loss to the universe" (104). Crucially, these remarks suggest that a lack of fellow feeling towards such obscure species might in itself be useful, allowing for a response to emerge that is less anthropocentric—not oriented towards personal feelings of loss but towards a wider sense of ecological injustice. Here, the absence of certain criteria for valuing nonhuman entities—be it a sense of emotional connection, aesthetic appreciation, or a recognition of a creature's economic or social "good"—makes space for a deeper response to emerge.

Resaint's realisation is useful for thinking about how attending to the loss of species that we feel ourselves to be distant from—both geographically and emotionally—might prompt readers to think and feel differently about extinction. Lorimer defines nonhuman charisma as a "relational property," whereby the distinguishing features of a creature determine its perception and subsequent evaluation (915). But what if relation is not possible, let alone easy, or if distinguishing features are lacking? What might a non-relational response to extinction look like? Our wager in this collection is that attending to literary and cultural representations of aquatic species in particular, which make up a vast percentage of "distant, unmet things" (Giggs 2020, 24) on the planet, might help to angle our thoughts past arguments that hinge on relationality as a route to ethical valuation, and away from received structures of feeling ("wouldn't we be sad to lose this little fellow?" 104). Insofar as the extinction of the vast majority of "blue" species—whales and dolphins notwithstanding—challenges our capacity to care, or to *feel blue*, in the way that we would about more familiar, terrestrial species, this absence of feeling may in turn provide opportunities to conceive of the loss of nonhuman species aside from questions of personal connection, social good, or economic gain.

Through its satirising of the mass mourning of Chiu Chiu, or the cynical monetisation of biodiversity loss by large-scale corporations, Beauman's narrative demonstrates how literature and other forms of artistic representation can expose and challenge entrenched cultural attitudes towards biodiversity loss. The novel demonstrates the ease with which narratives surrounding extinction can become mired in systems of valuation—be it ideas of nonhuman charisma or insidious notions of natural capital—that continue to privilege the lives of certain species over others ("Natural Capital" n.d.). That said, *Venomous Lumpsucker* is not entirely free of a tendency to conflate the worth of species with their instrumental value. By centring the novel around a marine species which not only displays a remarkable intelligence, but also performs an important ecological function as a prolific cleaner fish (or submarine hygienist), the text indicates that endangered organisms still need to demonstrate that they are worthy of preservation.

By focusing on a specific, albeit fictional, species of fish, moreover, the novel perpetuates a narrative of singularity within extinction discourse, in which the privileging of individual species or individualised members of a species (also known as "endlings") can serve to obscure more complex and collective ways of being (Pyne 2023). The venomous lumpsucker bears a striking resemblance to the poster child of the comedic conservation initiative "The Ugly Animal Preservation Society" n.d. Established in 2012 with the aim of "rais[ing] the profile of some of nature's more aesthetically challenged creatures," the Society is fronted by the blobfish, a creature which, the website informs us, was once voted the world's ugliest animal ("The Ugly Animal Preservation Society" n.d.). Like the blobfish, the venomous lumpsucker trades in a form of negative charisma, with its value remaining bound up with the strong affective response it elicits in the viewer. It would be difficult to conceive of a novel centred around a species of microscopic plankton, for instance, whose imperceptible, pelagic, deindividuated state eludes narrative coherence. A focus on individual experience, is, of course, the purview of the novel, and yet it may also be a generic limitation when it comes to representing forms of biodiversity loss which defy notions of singularity, and which, as the text rightly notes, involve not only a set of bodies, but a "a set of habits, relationships, territories, entanglements" (28).

As Ben Walker points out in his review of the novel, "for the sake of narrative ease ... it's not collective 'biological systems' that have financial value, but rather individual species with prices on their heads" (Walker 2023, n.p.). The notion of "narrative ease" implies that biological systems are too complex, too diffuse for this kind of storytelling. This is particularly the case when it comes to aquatic environments where there are so many interconnected, circulating environmental features and forces (including salinity, oxygenation, and toxicity), minor alterations to which can affect species en masse and in ways that defy terrestrial notions of boundedness, separation, and individuation. In bodies of water, it is difficult to isolate or contain many species, which is why marine protection initiatives tend to focus on protecting areas rather than species.[1] The depth and inaccessibility of the world's waters, moreover, makes it impossible to determine the full extent and scale of aquatic biodiversity loss. As one recent study notes, because somewhere between 75 and 90% of marine species are still yet to be classified, "estimating the loss of marine

[1] See for instance IUCN—Marine Protected Areas—Why Have Them? n.d. https://web.archive.org/web/20120319150803/http://www.iucn.org/about/work/programmes/pa/pa_what/?4646/Marine-Protected-Areas%2D%2DWhy-have-them/. Accessed March 26, 2024.

biodiversity is as much about what we do not know as what we do know" ("Habitat and Biodiversity Loss" n.d.). Yet Beauman's novel envisages a scenario where the DNA of every extinct species, blue or otherwise, has been captured and stored in a database. Though the biobanks are hacked, and the data is lost, the text fantasizes about a state of informational totality surrounding endangered species that remains practically impossible, especially at sea. As well as offering an incisive, and darkly funny vision of the marketisation of biodiversity loss, *Venomous Lumpsucker* presents some significant aesthetic, epistemic, and ethical challenges in representing blue extinction that this collection sets out to explore in further depth.

Framing Aquatic Biodiversity Loss

The extinction of marine and freshwater species has occurred throughout evolutionary history. Marine invertebrates bore the brunt of earlier mass extinctions, from the Late Ordovician to the Late Permian extinction event, in the latter of which it is estimated that up to 95% of marine species were wiped out compared to 75% of terrestrial organisms (Shen and Bowring 2014). Aquatic species have also been central to evolving scientific conceptions of biodiversity loss. In 1808, the French naturalist Georges Cuvier, who established the concept of extinction, used the fossil of a marine reptile known as the *Mosasaur* to support his argument that animals once lived on earth which no longer existed (Kolbert 2014, 27). Cuvier's theory of extinction as a sudden catastrophic event was widely disputed by leading nineteenth-century scientists such as Charles Darwin, who argued that extinction was a gradual process as well as a normal and necessary part of evolution. Like Cuvier, though, Darwin would base his theory on the study of marine fossils collected during his famous Beagle expedition (Stott 2004). In 1980, Cuvier's catastrophic hypothesis was to a certain extent confirmed, again with the help of an aquatic species. The sudden absence of ammonites—a small marine mollusc—from the fossil record had long perplexed evolutionary theorists such as Darwin, as well as the geologist Charles Lyell, who could not account for their "wonderfully sudden" disappearance (Kolbert 2014: 54). While studying sediments from limestone cliffs, however, father and son Walter and Luis Alvarez discovered evidence pointing to the mass extinction of dinosaurs and many other species resulting from a large asteroid. The timing of this late-Cretaceous extinction event has recently been pinpointed more accurately by scientists studying fossilised fish bones (Barras 2022), suggesting that marine species remain integral to the study of extinction.

In our present biodiversity crisis, aquatic species are once again proving themselves to be especially vulnerable to extinction. The Sixth Extinction, as it has come to be known, is resulting in the rapid loss of species at between 100 and 1000 times the normal background rate (de Vos et al. 2015). Recent studies suggest that marine and freshwater organisms are dying out at twice the rate of terrestrial species due to the combined effects of overfishing, pollution, climate change, water acidification, deep-sea mining, and other human impacts (Pinsky et al. 2019, 110). As we began writing this introduction in the summer of 2023, reports of a global marine heatwave were threatening to trigger more mass coral bleaching events, high rates of fish mortality, and the collapse of the Gulf Stream. Add to this the looming prospect of deep-sea mining by commercial organisations seeking to exploit natural resources such as cobalt and nickel, along with the widespread dumping of sewage by water companies and the failure of Marine Protected Areas (or MPAs) to protect against illegal harvesting, and the future of blue ecosystems appears especially bleak. Some scientists have estimated that if current greenhouse emissions continue, by 2100, as many as 90% of marine species will be extinct (Boyce and et al. 2022, 861). This, compared with 50% of terrestrial species (Dawson 2016, 9), confirms our belief that the importance of attending to blue—as well as green—biodiversity loss cannot be overstated.

As we have already seen, there are compelling reasons for drawing on literary, artistic, and historical approaches to blue extinction alongside, and in addition to, natural and social-scientific perspectives. A lack of scientific data due to the financial costs and practical challenges of deep-sea exploration makes aquatic biodiversity loss harder to quantify or visualise. Blue extinctions tend to occur out of sight and on a scale that it is difficult to fathom, meaning that imaginative responses become all the more important. Of course, many of the extinctions that are occurring on land are also largely invisible or too vast to comprehend—we might recall Resaint's parasitic wasp. Yet, as Timothy Clark points out: "The topic of the oceans highlights that major feature of the global environmental crisis, the disjunction between day-to-day perception (land-based, limited in scale) and broader decisive realities that are as out of sight as the ocean depths" (Clark 2019, 117). In the face of such empirical and epistemic limitations, other ways of knowing blue species need to be drawn on, and other modes of response engaged.

In a recent essay collection on the value of not knowing, the literary critic Emily Ogden presents the following epigraph from Herman Melville's *Moby-Dick* (1851) to articulate her approach: "Unfitness to pursue our research in the unfathomable waters" (Ogden 2022, n.p.). This statement, which features in the "Cetology" chapter of the novel, was lifted by Melville from a scientific paper written by the British anatomist John Hunter on the "structure and œconomy of whales" from 1787, which begins:

> The animals which inhabit the sea are much less known to us than those found upon land; and the œconomy of those with which we are best acquainted is much less understood: we are, therefore, too often obliged to reason from analogy where information fails; which must probably ever continue to be the case, from our unfitness to pursue our researches in the unfathomable water. (Hunter 1787, 371)

Hunter acknowledges his reliance on a literary and philosophical trope—analogy—to reason "where information fails." This passage long precedes the era of the "two cultures" debate, as outlined by C. P. Snow in the mid-twentieth century, where science and the humanities were thought to be entirely separate from one another (Snow 2012 [1959]). Yet it is helpful for thinking about what an approach anchored in the humanities might have to offer when it comes to grappling with the difficulties of aquatic biodiversity loss. Insofar as literature and art can help us to descend into waters that are largely unfathomable, and which can make us feel unfit to pursue our research, so can it help us to linger in the uncertainty of blue extinction without seeking to resolve its murky depths into brightly illuminated facts.

It is important to recognise, moreover, that literary texts such as *Moby-Dick* have functioned as a key source of cultural knowledge about aquatic biodiversity loss. As the environmental historian John Gillis notes, since the early nineteenth century "fiction has been imagining undersea worlds that explorers were unable to reach," meaning that "[w]e have come to know the sea as much through the humanities as through science" (Gillis 2013, n.p.) From Melville's cetaceans to Jules Verne's giant squid, mainstream knowledge of blue species has been shaped by often outlandish literary and artistic visions. Such imaginative explorations have even been drawn on by scientists: as Killian Quigley points out in his chapter, oceanographers and marine scientists continue to draw on passages from Jules

Verne's undersea adventure novel *Twenty Thousand Leagues Under the Seas* (1870) to forecast various nightmarish scenarios concerning ocean health. Rick De Vos's chapter, meanwhile, examines how Western cultural and scientific representations of narwhals as mythical sea-unicorns have combined to misrepresent these species in ways that have had a detrimental effect on their numbers. As Kathyrn Ferguson has elsewhere argued, because our engagement with aquatic species has largely been mediated through literature, art, and culture—from novels, to paintings, to nature documentaries—this has left them "more vulnerable than most to processes of misrepresentation and exploitation" (Ferguson 2006, 126). Who in the West can look at a shark without hearing the ominous notes from *Jaws*? According to Ursula Kluwick, while shark attacks are statistically rare, their predation by humans remains widespread; indeed, the popular imagining of the shark as hunter serves to elide the more disturbing reality of the shark as hunted to the point of endangerment (Kluwick 2015, 139).

As several chapters in this collection demonstrate, Western literature and art has played a major role in enshrining unhelpful and at times harmful blue cultural tropes. These include the myth of inexhaustibility, as expressed by the English biologist T. H. Huxley during his inaugural address of the Fisheries Exhibition in London in 1883: "all the great sea fisheries, are inexhaustible," he proclaimed, "that is to say that nothing we do seriously affects the number of the fish" (Khetan et al. 2023, 2162). This vision of oceanic ecosystems as a vast and limitless resource has been traced back as far as Aeschylus's *Oresteia* (c. 400 BC), when Clytemnestra states: "There is the sea—who shall exhaust the sea?" (Shepherd 1997, 843). Kimberley Patton has explored how ancient Greek representations of the sea as a purifying force that can "wash away all evils" has hampered the recognition of the effects of anthropogenic pollution on oceanic ecosystems (Patton 2006). Meanwhile, the long-standing cultural trope of the ocean as timeless, or beyond time, has come back to haunt us in recent years. As Astrida Neimanis argues, this mythic temporality fosters the false notion that "we can cast off our dumped desires to a time that is no longer now"—desires which take the form of "plastic polymers, fishing nets, nitrogen, phosphorus, radioactive isotopes," the list goes on (Neimanis 2019, 495). Even as sewage and other human detritus wash up on our shores, the ocean depths still harbour a fantasy that "we can relinquish what we no longer want, that our spent desires can be ferried away to a far away and long ago, that damage will never arrive" (Neimanis 2019, 495).

Literary and cultural works have also played a major role in constructing aquatic species as figures of fascination and horror. From nineteenth-century adventure fiction to contemporary science fiction films, mainstream Western culture abounds with representations of ocean depths filled with mythical monsters and "alien" sea beasts (Helmreich 2009). As Stacy Alaimo has noted elsewhere, the popular trope of marine life as "alien" hampers conservation efforts by imaginatively removing these species "from the terrain of human concern, and even from reality" (Alaimo 2017, 154). Such fantastical constructions of blue species present a vision of the undersea world as untouched by human activity, and as a space "free of guilt from histories of colonialist oppression and ecological harm" (Alaimo 2020, 317). This settler-colonial optic became particularly apparent during the advent of underwater film in the early twentieth century, when the ocean depths were "discovered" and brought into view by white American and European filmmakers (Elias 2019). As Nicole Starosielski has argued, the works of the pioneering American underwater filmmaker John Ernest Williamson enact a kind of double vision, in which depictions of violence against marine animals and the exploitation of Black Indigenous subjects coexist alongside descriptions of the ocean depths as an untouched and pristine wilderness (Starosielski 2012). This way of seeing is sadly not confined to early underwater film or to the early twentieth century. As Helen F. Wilson has noted, while the recent BBC nature documentary series *Blue Planet II* (2017) served to raise awareness of anthropogenic harm to oceanic species and galvanise public conservation efforts, it also continues to present a Western imperialist vision of the undersea as a "final frontier on earth" and a "vast, empty blue" (Wilson 2019, 719). By mourning the loss of a so-called pristine underwater nature, the series upholds the very power structures that it purports to critique. To continue to delve into these blue archives, then, is vital to recovering and extending our historical awareness of aquatic biodiversity loss as well as its ongoing legacy.

Approaching the "Blue"

This collection draws on existing work in the blue humanities, which has sought to challenge the terrestrial bias of environmental thought, and to reorient critical attention offshore and towards inland bodies of water (Blum 2010; Mentz 2020; Oppermann 2023). Our approach takes its cues from scholarship which has highlighted the importance of focusing on water in its "twofold condition" (Oppermann 2019, 451) as both a material and a metaphorical construct, coloured by centuries of symbolism as well as synthetic substances. It also seeks to build on the transnational, global aspect of the blue humanities, foregrounding perspectives from the Global South, as well as Black and Indigenous literature and art (DeLoughrey 2020; Hofmeyr 2019; Lavery 2021; Samuelson 2017; Shewry 2015; Santos Perez 2020). While several chapters in this collection draw on scientific knowledge of marine and freshwater ecosystems to comprehend the causes and effects of blue extinction events, our contributors also seek to interrogate the epistemological dominance of Western scientific thought, taking the lead from Black feminist thinkers such as Alexis Pauline Gumbs, who have strived to make room for less dominant, and more constructive ways of knowing and relating to aquatic species (Gumbs 2021).

This collection also builds on existing work in the blue humanities by presenting a sustained focus on animals and extinction. While the subject of aquatic biodiversity loss has been touched on in several studies (Alaimo 2013; Jue 2020; Huggan 2018), and while critical attention has been paid to specific species such as corals, frogs, fish, and lobsters (Schuster 2019; Boast 2022; Todd 2017; Johnson 2015), so far there has been no sustained focus on marine and freshwater extinction that has done justice to its scale, severity, or variety. Our sense, moreover, is that the abundance of watery metaphors that have characterised some blue humanities scholarship—including ideas of fluidity, flow, mobility, salinity, dynamism, drowning, depth—have at times served to shift attention away from the specific material, political and cultural factors that are impinging on oceanic and freshwater ecosystems (Oppermann 2019). While continuing to "think with water" (Chen et al., 2013), and its infiltration by pollution and microplastics (Bloomfield 2019; Jaegar 2010), this collection seeks to shift the focus from bodies of water to the organisms that inhabit them and the threats they face. While maintaining a large-scale perspective on nonhuman species, environmental ethics, and multispecies justice, this collection homes in on aquatic extinction as a necessary and urgent topic.

Several of our contributors seek to challenge some of the more utopian impulses surrounding the "blue," which can be in danger of framing aquatic environments and ways of being as an emancipatory alternative to terrestrial existence. As O'Key and Akshita Bhardwaj argue in their chapter: "An 'oceanic turn' … will be for nothing if it means turning away from tricky terrestrial relations" or "fetishizing or romanticizing the otherness of the ocean" (see Chap. 2). Positioning the blue humanities as a turn away from the green runs the risk of formulating waters as a restorative force, or antidote to capitalist and colonialist exploitation on land, thus falling back into the misleading cultural trope of the sea as an agent of purification and overlooking its infiltration by terrestrial power structures. In their chapter, Ivry and Karpinski coin the helpful term "bluewashing" to describe the inclusion of the work of Black thinkers such as Du Bois in oceanic scholarship without acknowledging the racialised violence that undergirds their ideas. Such a glaring omission, they suggest, should cause us to reflect more broadly on what cannot be left out of any critical consideration of the ocean, and what a romanticisation of the blue might seek to wash away.

By approaching aquatic environments through the lens of anthropogenic extinction, our collection seeks to tie these watery spaces directly to the realities of Western environmental destruction and the ongoing effects of imperial and racial violence. While maintaining a focus on the blue, our contributors have a marked interest in the limitations, as well as the possibilities of this term. Rather than confining its focus to an isolated blue sphere, our hope is that this collection will open up conversations about the health and wellbeing of aquatic biodiversity in conjunction with terrestrial ecosystems and power structures. In addition to unsettling the boundary between the blue and the green, many of our chapters highlight the porosity of the distinction between near and distant environments and organisms. By focusing on lesser-known aquatic species residing in the remote ocean depths, several of our contributors seek to examine how environmental destruction occurring in nearby waters, be it physical (in the forms of pollution and habitat destruction) or political (in the forms of legislation and its enforcement) impacts distant waters, and vice versa.

"Storying" Blue Extinction

The chapters in this collection bring the blue humanities into productive dialogue with extinction studies, a field of research that examines the literary, cultural, social, and political aspects of biodiversity loss (Heise 2016; Rose et al. 2017). Extinction studies is most often associated with the Extinction Studies Working Group, a collective of humanities scholars who are "interested in exploring multi-disciplinary approaches to extinction" (About us "Extinction Studies Working Group" n.d.). The work of this group of scholars, who include anthropologists, environmental philosophers, and critical theorists, centres around practices of storytelling (or storying), field work, direct encounter, and ethical response. Members have sought to emphasise extinction as a cultural, as well as a biological phenomenon "where not just individual organisms, but entire ways and forms of life, are at stake" (Rose et al. 2017, 5). Our collection draws on various aspects of this approach, including interdisciplinarity, fieldwork, ethical response, and multispecies thinking. Our focus on aquatic biodiversity loss is particularly indebted to the recent work of the extinction studies of scholars Michelle Bastian and Thom van Dooren, who have sought to move the emphasis away from direct encounters with known species, or last known members of a species (endlings), towards storying "unknown" extinctions (Bastian 2020; Van Dooren 2022). In an article on whale falls, Bastian encourages readers to contemplate the invisible deep-sea extinctions that may have taken place following reduced numbers of "falling" whales due to industrial hunting (456). For Bastian, focusing on the invisible losses of deep-sea species that rely on whale falls has powerful ethical implications insofar as it encourages a tracing of one's own largely invisible "connections with, and responsibilities for, the processes causing these losses." (471).

Bastian's article has been crucial to our understanding of how aquatic species pose a set of productive perceptual, epistemological, and ethical challenges for apprehending extinction, shifting the focus from apprehensible losses to larger unseen processes. With the exception of Bastian's work, however, the emphasis in extinction studies has largely been on terrestrial animals such as parakeets, bison, dogs, and snails.[2] Our wager is that a more sustained focus on blue species, as well as prompting greater recognition of the value of tarrying with unknown or poorly understood creatures, may help to loosen the grip of certain conservation ideals and affective attachments that shadow "storying"' practices, including physical proximity, direct and/or sustained encounter, charisma, intelligence, and economic value. This includes negative charisma, which as we saw earlier in the case of the "ugly" venomous lumpsucker, can still hold attention and trigger curiosity in ways that privilege aesthetic appeal and singularity over obscurity and collectivity. The importance of focusing on uncharismatic, overlooked, and imperceptible species is, in our view, vital in contemplating less obvious, yet potentially profound, responses to nonhuman others—including those who play more crucial ecosystem roles than those

[2] See also Bastian's work on leatherback turtles (2017).

who have established their fame. This collection seeks to foreground the ethical importance of tarrying with creatures who cannot be fully known or directly encountered—what the marine biologist Helen Scales refers to as "hidden life" (Scales 2021, 35)—which challenge epistemological certainty and serve as a powerful reminder of certain invisible lines of complicity. How might it be possible to cultivate, and sustain, an ethical sensitivity towards that which remains beyond our ken? This collection invites readers to dwell imaginatively, and affectively, in submerged ecosystems impacted by various forms of darkness.

Within extinction studies, the practice of "storying" biodiversity loss has become widespread. As Dominic O'Key notes elsewhere, the primary aim of this approach is to "mobilize readers' emotions" through acts of witnessing, often homing in on a particular endangered species, lamenting its demise, and analysing the wider ethical implications of its loss from the world (O'Key 2023, 179). Yet blue extinction, with its complex blend of systemic interactions, does not easily lend itself to storying the "intimate particulars" of a species or its ways of life (Bird Rose, qtd by Bastian, 464). Given the difficulties of encountering endangered aquatic species—even fleetingly—coupled with the fact that so many species remain undiscovered in life, let alone in death, this process of affective mobilisation is especially difficult. In aquatic environments, organisms often exist either in groups or can only be known collectively as assemblages, examples of which include coral, osedax worms, or zooplankton. This is also reinforced by collective nouns pertaining to aquatic animals—widely used in research, industry, and tourism—such as pods of whales or dolphins, floats of tuna, shoals of mackerel, runs of salmon, families of sardines, herds of sea urchins. Blue extinction can occur in relation to an entire species (global extinction), but also to a group living in a particular environment (local extinction); the latter can lead to the global extinction of dependent species. To focus on blue extinction is therefore to focus on extinctions unknown and unfelt par excellence, comprised as they are of losses that tend to be distant, multiple, indistinct, and poorly understood.

As Dolly Jørgensen points out, extinction already poses a challenge to conventional storytelling due to its focus on complex processes rather than singular events, and because it "includes simultaneously deep-time elements and rapid change" (Jørgensen 2022, 210). The notion of storying can feel particularly constraining when it comes to representing blue extinction, often focused as it is on individual species, on personal feelings of loss ("wouldn't we be sad to lose this little fellow?" 104), and on acts of witnessing and direct encounter. As Stacy Alaimo argues in her chapter, a practice of storying may not be possible, or appropriate, for representing "the vast majority of undiscovered species who have been rendered extinct by human activities." Rather, "aesthetic encounters" may be preferable to narrativisation in engaging with "beings that resist human stories" (see Chap. 10). At the same time, the representational challenge posed by blue extinction may put productive pressure on the act of storying. Several of our contributors show how blue extinction can at once expose the inefficacy of traditional narrative methods while occasioning new and innovative approaches. In her chapter on cephalopodic consciousness, Rosanne van der Voet experiments with language and typography to reflect on the challenges, as well as the possibilities, of storying the "multi-limbed mind" (see Chap. 9). And in his chapter on oceanic violence and intimacy, Mentz shifts from academic prose into "poetic fantasy" to represent the impossibility of "full intimacy" between humans and whales: "[t]he poem imagines, but cannot quite present, a world big enough for all" (see Chap. 5). Serpil Oppermann has argued that "diverse aqueous life-forms call for our attentiveness to their stories that run alongside human narratives, transmitting frustrating messages of ecological despoilation [sic]" (2023, 4). The key word here is frustrating, which suggests both thwarted progress and a feeling of irritation. Focusing on the various frustrations—including generic and formal—of blue extinction can both expand and nuance our understanding of the spatial, temporal, and epistemological complexities of aquatic biodiversity loss along with its varied cultural manifestations.

The chapters in this collection explore the subject of blue extinction across a wide range literary and artistic forms, including novels and poetry, music, art installations, museum exhibits, photography, film, and visual media. Traversing the early modern period to the present and moving from violent whaling histories to slimy oceanic futures, our contributors carry us from the littoral through to the abyssal zones, and from warm Bengali waters to Antarctic ice sheets. The subject matter of these chapters ranges from unmissable to imperceptible forms of aquatic life—from the vastness of endangered whales to the humble sea mouse (a type of marine worm). As Vera Fibisan notes in her chapter, many marine invertebrates "run the risk of falling between the cracks of conservation narratives" despite playing such "important roles in intertidal ecosystems" (see Chap. 8). Several of our contributors seek to remedy this critical neglect of aquatic invertebrates and other non-mammalian life forms, tracing literary and cultural responses to crabs, oysters, and coelacanths (an ancient deep-sea fish) in addition to whales, seals, and dolphins.

The opening chapters in this collection focus on the challenge of representing aquatic biodiversity loss. Akshita Bhardwaj and Dominic O'Key's chapter focuses on the figure of the dolphin in Amitav Ghosh's *The Hungry Tide* (2004) and its follow-up, *Gun Island* (2019), tracing this vulnerable cetacean through two novels which are commonly read and analysed from the perspective of climate fiction. Arguing that Ghosh's dolphins push back against Americanised portrayals of dolphins as cheerful, intelligent, and charismatic figures, Bhardwaj and O'Key show how these novels nonetheless run the risk of presenting dolphins as salvific creatures, "romanticising" the blue as an alternative to the fraught terrain of "green imperialism" (see Chap. 2). Killian Quigley's chapter continues the theme of figuration by examining how contemporary oceanographers and marine scientists have drawn on representations of extinctive future seas in Jules Verne's submarine adventure novel *Twenty Thousand Leagues Under the Seas* (1870). While recognising the prescience of Verne's vision of "slimy" seas "cluttered with octopi, medusae, and squids," Quigley also reflects critically on the origins of this oceanic prediction, which lie in the misanthropic and xenophobic writing of the French naturalist Alphonse Toussenel. The "literariness of recent attempts to figure ocean futures in general, and marine extinctions in particular," Quigley tells us, "has histories. Those histories make available cautionary information respecting the genealogies of certain narratives, affects, and imageries" (see Chap. 3).

In Chap. 4, Rick De Vos focuses on the "elusive" figure of the narwhal, a species that Western science has had a hard time observing and studying, in part, because it spends half of the year living beneath remote Arctic sea-ice. In contrast to Indigenous cultural representations, which result from direct contact with, and knowledge of narwhals, De Vos examines how the dearth of scientific information about narwhals has resulted in Western cultural representations—including children's picture books, museum displays, and toys—that move them "into the realm of fantasy and magic." As with the previous two chapters, De Vos highlights the dangers that come with "misrepresenting the agency, complexity, and diversity" of aquatic species (see Chap. 4).

The chapters that follow share an emphasis on alternative affective responses to aquatic species. Steve Mentz traces the historical movement of whales in Western culture from desired commodities to charismatic species fronting conservation campaigns, before arguing that a dual emphasis on violence and vulnerability can be traced all the way back to the early modern accounts of whaling. Inspired by Gumbs's "radical empathy" towards marine mammals in her recent book *Undrowned: Black Feminist Lessons from Marine Mammals* (2020), Mentz analyses a series of "whale encounters"—both imaginary and real—in literary writing from Shakespeare to the present "to suggest how these creatures can expand our ecological thinking, inspire oceanic intimacies, and re-order ideas of the human" (see Chap. 5). Megan Hayes's chapter considers the biological and affective modes of attachment modelled by the humble oyster—a species whose industrious filtering processes contribute to the health of marine ecosystems worldwide. Exploring the problematics of oyster restoration initiatives such as the Billion Oyster Project (BOP) and the Living Breakwaters project, Hayes's chapter proposes that such "gray and green infrastructures" are serving to distract from more pressing problems faced by aquatic environments as well as Indigenous peoples. Exploring the similarities between oysters and humans, while drawing on the latter's adaptive abilities, Hayes advocates for "hope for ecological salvation" through reparation via "a molluscan ethics of entanglement" (see Chap. 6).

A focus on embodied encounters with aquatic species animates the chapters that follow. Karen Eckersley's exploration of Elizabeth Bishop's neglected triptych of prose poems, *Rainy Season; Sub-Tropics* (1969), which focuses on a trio of aquatic creatures (a giant snail, a strayed crab, and a giant toad), argues that Bishop utilises the form of the prose poem to "immerse readers in a poetic and virtual seascape of slippages and movement." Like the creatures themselves, the reader is encouraged to "slide between invocations of land and water … species and environment, ancient and current temporalities" (see Chap. 7), as they navigate this hybrid poetic sequence. In the next chapter, Vera Fibisan examines Elizabeth-Jane Burnett's poetry collection *Of Sea* (2021), arguing that her practice of "swimwork," coupled with her lyric experimentation, evokes a series of intimate encounters with aquatic invertebrates under the threat of extinction. Drawing on Neimanis's concept of "membrane logic," Fibisan explores how Burnett's hydromaterialist poetics establishes "more-than-human bonds between species, temporalities, and substances 'in the twirling muscle / of the present / & the scream of being / one thing' (36)" (see Chap. 8).

In the final chapters, we descend into the ocean depths to consider the submerged presence of "violet-black" (Alaimo 2013) ecologies and "Black *and* blue" (see Chap. 11)) extinctions. Rosanne van der Voet's creative-critical chapter, "A Story of Eight Limbs," explores the challenges of storying deep-sea octopi at a time of extinction, stretching the limits of human storytelling in her creative exploration of cephalopodic agency. In dialogue with Oppermann's concept of "storied seas" (Oppermann 2019), van der Voet's chapter experiments with layout, typography, and textual opacity to develop a form of language divested of its anthropocentric moorings: "Slowly diffusing their human meaning, [words] spill into the margins of this text, swelling with strange and unexpected intimacies" (see Chap. 9). Stacy Alaimo's chapter then delves into the "abyssal ecologies" of Anthropocene seas, arguing that the uncanny figuration of "enduring 'living fossils'" such as coelacanths on the one hand, and "presumed but unknown abyssal extinctions" on the other, can be mobilised against the threat of deep-sea mining. "Spectral species of the abyss," Alaimo surmises, "as oxymorons, mind-warping temporalities, speculative unknowability, and strangely fragile solidity, call us to navigate the turbulent places where multiple temporalities affect ethical and political agency" (see Chap. 10). The collection concludes with a set of subaquatic soundings by Henry Ivry and Max Karpinski, who tune in to the disruptive "alien rhythms" of Claude McKay's modernist port novel *Banjo* (1929) and the electronic music of Drexciya. "Black artists in the diaspora," they argue, "have been sounding out a nascent oceanic ecology that calls into presence alternative figurations of Black futurities" that unsettle intelligibility and defy the "killing rhythm" of capitalism and colonialism (see Chap. 11).

There remains further work to be done in uncovering Black and Indigenous, as well as other marginalised ways of knowing and relating to aquatic species in a time of biodiversity loss—including queer and trans ecological perspectives. There is also more to be said about literary and cultural representations of freshwater as well as marine species, and we note that charismatic megafauna such as whales and dolphins still take up a great deal of space in the collection. The role of Western literature and culture in producing and disseminating knowledge of aquatic animals and ecosystems—either in conjunction with, or in opposition to, scientific accounts—also warrants further investigation. But our hope is that this collection, while not exhaustive, will open up rich avenues of future research on the subject of blue extinction across a wide range of literary and cultural forms.

References

About Us. n.d. *Extinction Studies Working Group*. https://extinctionstudies.org/about/. Accessed 26 March 2024.

Alaimo, Stacy. 2013. Violet-Black. In *Prismatic Ecologies: Ecotheory Beyond the Green*, ed. Jeffrey Jerome Cohen and Lawrence Buell, 233–251. Minneapolis: University of Minnesota Press.

———. 2017. The Anthropocene at Sea: Temporality, Paradox, Compression. In *The Routledge Companion to the Environmental Humanities*, ed. Ursula Heise, Jon Christensen, and Michelle Niemann, 153–161. London and New York: Routledge.

———. 2020. Adequate Imaginaries for Anthropocene Seas. In *Blue Legalities: The Life and Laws of the Seas*, ed. Irus Braverman and Elizabeth Johnson, 311–326. Durham, NC: Duke University Press.

Barras, Colin. 2022. Fossil Fish Reveal Timing of Asteroid that Killed Dinosaurs. *Nature* 23. https://www.nature.com/articles/d41586-022-00511-x. Accessed 26 March 2024.

Bastian, Michelle. 2017. Encountering Leatherbacks in Multispecies Knots of Time. In *Extinction Studies: Stories of Time, Death and Generations*, ed. Deborah Bird Rose, Thom van Dooren, and Matthew Chrulew, 149–186. New York: Columbia University Press.

———. 2020. Whale Falls, Suspended Ground, and Extinctions Never Known. *Environmental Humanities* 12 (2): 454–474.

Beauman, Ned. 2022. *Venomous Lumpsucker*. London: Sceptre.

Bloomfield, Mandy. 2019. Widening Gyre: A Poetics of Ocean Plastics. *Configurations* 27 (4): 501–523.

Blum, Hester. 2010. The Prospect of Oceanic Studies. *PMLA* 125 (3): 670–677.

Boast, Hannah. 2022. Theorizing the Gay Frog. *Environmental Humanities* 14 (3): 661–679.

Boyce, Daniel G., et al. 2022. A Climate Risk Index for Marine Life. *Nature* 12: 854–862.

Chen, Celia, Astrida Neimanis, and Janine MacLeod, eds. 2013. *Thinking with Water*. Montreal: McGill-Queen's University Press.

Clark, Timothy. 2019. *The Value of Ecocriticism*. Cambridge: Cambridge University Press.

Dawson, Ashley. 2016. *Extinction: A Radical History*. New York: OR Books.

DeLoughrey, Elizabeth. 2020. *Allegories of the Anthropocene*. Durham, NC: Duke University Press.

de Vos, Jurriaan M., Lucas N. Joppa, John L. Gittleman, Patrick R. Stephens, and Stuart L. Pimm. 2015. Estimating the Normal Background Rate of Species Extinction. *Conservation Biology* 29 (2): 452–462. https://doi.org/10.1111/cobi.12380.

Elias, Ann. 2019. *Coral Empire: Underwater Oceans, Colonial Tropics, Visual Modernity*. Durham, NC: Duke University Press.

Ferguson, Kathryn. 2006. Submerged Realities: Shark Documentaries at Depth. *Atenea* 26 (1): 115–129.

Giggs, Rebecca. 2020. *Fathoms: The World in the Whale*. New York: Simon and Schuster.

Gillis, John. 2013. The Blue Humanities. *NEH*. https://www.neh.gov/humanities/2013/mayjune/feature/the-blue-humanities. Accessed 26 March 2024.

Gumbs, Alexis Pauline. 2021. *Undrowned: Black Feminist Lessons from Marine Mammals*. Chico, CA: AK Press.

Habitat and Biodiversity Loss. n.d. *Our Shared Seas*. https://oursharedseas.com/threats/threats-habitat-and-biodiversity/#n15. Accessed 26 March 2024.

Heise, Ursula K. 2016. *Imagining Extinction: The Cultural Lives of Endangered Species*. Chicago: University of Chicago Press.

Helmreich, Stefan. 2009. *Alien Ocean: Anthropological Voyages in Microbial Seas*. Berkeley, CA: University of California Press.

Hofmeyr, Isabel. 2019. Literary Ecologies of the Indian Ocean. *English Studies in Africa* 62 (1): 1–7.

Huggan, Graham. 2018. *Colonialism, Culture, Whales: The Cetacean Quartet*. London: Bloomsbury Academic.

Hunter, John. 1787. Observations on the Structure and Economy of Whales. *Philosophical Transactions of the Royal Society of London* 77 (7): 371–451. https://royalsocietypublishing.org/doi/10.1098/rstl.1787.0038.

Jaegar, Patricia. 2010. Editor's Column: Sea Trash, Dark Pools, and the Tragedy of the Commons. *PMLA* 125 (3): 523–545.

Johnson, Elizabeth. 2015. Of Lobsters, Laboratories, and War: Animal Studies and the Temporality of More-Than-Human Encounters. *Environment and Planning D: Society and Space* 33 (2): 296–313.

Jørgensen, Dolly. 2022. Extinction and The End of Futures. *History and Theory* 61 (2): 209–218.

Jue, Melody. 2020. *Wild Blue Media: Thinking through Seawater*. Durham, NC: Duke University Press.

Khetan, Aditya, Vittal Hejaji, and Richard Josephson. 2023. Are We Running Out of Fish? Fish, Health and Sustainability. *Public Health Nutrition* 26 (10): 2162–2164.

Kluwick, Ursula. 2015. Food for Sharks: Abjection on the Beach. In *The Beach in Anglophone Literature and Cultures: Reading Littoral Space*, ed. Ursula Kluwick and Virginia Richter, 139–154. London: Routledge.

Kolbert, Elizabeth. 2014. *The Sixth Extinction: An Unnatural History*. New York: Henry Holt and Co.

Lavery, Charne. 2021. *Writing Ocean Worlds: Indian Ocean Fiction in English*. Cham, Switzerland: Palgrave Macmillan.

Lorimer, Jamie. 2007. Nonhuman Charisma. *Environment and Planning D: Society and Space* 25 (5): 911–932.

Mentz, Steve. 2020. *Ocean: Object Lessons.* London and New York: Bloomsbury Academic.

Natural Capital and Ecosystem Services. n.d. *United Nations.* https://seea.un.org/content/natural-capital-and-ecosystem-services-faq. Accessed 26 March 2024.

Neimanis, Astrida. 2019. The Weather Underwater: Blackness, White Feminism, and the Breathless Sea. *Australian Feminist Studies* 34 (102): 490–508.

O'Key, Dominic. 2023. Extinction in Public: Thinking through the Sixth Mass Extinction, Environmental Humanities, and Extinction Studies. *Environmental Humanities* 15 (1): 168–186.

Ogden, Emily. 2022. *On Not Knowing: How to Love and Other Essays.* Chicago: University of Chicago Press.

Oppermann, Serpil. 2019. Storied Seas and Living Metaphors in the Blue Humanities. *Configurations* 27 (4): 443–461.

———. 2023. *Blue Humanities: Storied Waterscapes in the Anthropocene.* Cambridge: Cambridge University Press.

Patton, Kimberly. 2006. *The Sea Can Wash Away All Evils: Modern Marine Pollution and the Ancient Cathartic Ocean.* New York: Columbia University Press.

Pinsky, Malin L., Anne Maria Eikeset, Douglas J. McCauley, Jonathan L. Payne, and Jennifer M. Sunday. 2019. Greater Vulnerability to Warming of Marine versus Terrestrial Ectotherms. *Nature* 569: 108–111.

Pyne, Lydia. 2023. *Endlings: Fables for the Anthropocene.* Minneapolis: University of Minnesota Press.

Rose, Deborah Bird, Thom van Dooren, and Matthew Chrulew. 2017. *Extinction Studies: Stories of Time, Death and Generations.* New York: Columbia University Press.

Samuelson, Meg. 2017. Coastal Form: Amphibian Positions, Wider Worlds, and Planetary Horizons on the African Indian Ocean Littoral. *Comparative Literature* 69 (1): 16–24.

Santos Perez, Craig. 2020. "The Ocean in Us": Navigating the Blue Humanities and Diasporic Chamoru Poetry. *Humanities* 9 (66): 1–11.

Scales, Helen. 2021. *The Brilliant Abyss: True Tales of Exploring the Deep Sea, Discovering Hidden Life and Selling the Seabed.* London: Bloomsbury.

Schuster, Joshua. 2019. Coral Cultures in the Anthropocene. *Cultural Studies Review* 25 (1): 85–102.

Shen, Shu-zhong, and Samuel A. Bowring. 2014. The End Permian Mass Extinction: A Still Unexplained Catastrophe. *National Science Review* 1 (4): 492–495.

Shepherd, John. 1997. Oceans of Truth. *Nature* 388. https://www.nature.com/articles/42184. Accessed 26 March 2024.

Shewry, Teresa. 2015. *Hope at Sea: Possible Ecologies in Oceanic Literature.* Minneapolis: University of Minnesota Press.

Snow, C. P. 2012 [1959]. *The Two Cultures.* New York: Cambridge University Press.

Starosielski, Nicole. 2012. Beyond Fluidity: A Cultural History of Cinema under Water. In *Ecocinema: Theory and Practice*, ed. Stephen Rust, Salma Monani, and Sean Cubitt, 149–168. New York: Routledge.

Stott, Rebecca. 2004. *Darwin and the Barnacle.* London: Faber and Faber.

The Ugly Animal Preservation Society. n.d.. https://uglyanimalsoc.com/. Accessed 26 March 2024.

Todd, Zoe. 2017. Fish, Kin and Hope: Tending to Water Violations in the *amiskwaciwâskahikan* and Treaty Six Territory. *Afterall: A Journal of Art, Context, and Enquiry* 43 (1): 102–107.

van Dooren, Thom. 2022. In Search of Lost Snails: Storying Unknown Extinction. *Environmental Humanities* 14 (1): 89–109.

Walker, Ben. 2023. Kinda Wispy. *London Review of Books* 45 (3) https://www.lrb.co.uk/the-paper/v45/n03/ben-walker/kinda-wispy. Accessed 26 March 2024.

Wilson, Helen F. 2019. Contact Zones: Multispecies Scholarship through Imperial Eyes. *Environment and Planning E: Nature and Space* 2 (4): 712–731.

CHAPTER 2

Amitav Ghosh's Dolphins: Extinction, Figuration and Redemption in *The Hungry Tide* and *Gun Island*

Akshita Bhardwaj and Dominic O'Key

The Ocean'd Earth: Amitav Ghosh and Blue Extinction

Over the past decade the Brooklyn-based Indian author Amitav Ghosh has accrued new accolades and audiences for his creative-critical writings on the relationship between literature and climate. Ghosh, one of the foremost figures in contemporary world literature, is an author with an acute scientific and historical awareness. A trained anthropologist who builds his novels out of periods of research, he has been described as balancing "a global, ecumenical perspective" with stories of "highly individual, often contested and marginalized histories" (Sankaran 2012, xv). To mediate

A. Bhardwaj
English and Foreign Languages University, Hyderabad, India

D. O'Key (✉)
University of Sheffield, Sheffield, UK
e-mail: d.okey@sheffield.ac.uk

© The Author(s), under exclusive license to Springer Nature Switzerland AG 2024

V. Fibisan, R. Murray (eds.), *Blue Extinction in Literature, Art, and Culture*, Palgrave Studies in Animals and Literature,
https://doi.org/10.1007/978-3-031-69910-8_2

between these two contrasting scales of the planetary and the social, Ghosh often looks to the thematics of the blue. Indeed, across different forms of writing—fiction, essay, travelogue—Ghosh has long been preoccupied with the Indian Ocean, coastal communities and colonial maritime conquests. In more recent works he has developed this interest by turning much more explicitly towards the environment. In *The Hungry Tide* (2005) and its loose follow-up novel *Gun Island* (2020), Ghosh's narration grapples with how the strange reality of climate change problematizes the world-making mimesis of literary realism, asking: if the novel is an aesthetic form birthed out of a disenchanted modernity, then to what extent can it describe and plot the uncanny return of an enchanted nature, of "storms, floods, and unusual weather events" (Ghosh 2016, 15)?

With this question in mind, it is understandable that literary critics have assimilated Ghosh's writings into an ecocritical register. Drawing on aquatic metaphors like the pluvial and the diluvial, critics have explored how Ghosh navigates the problem of writing climate by dramatizing water as a world-remaking force (Anand 2008; Ganguly 2020; Kluwick 2020; Nuttall 2020; Wilton 2021). However, one risk of this critical move is that it might imprison Ghosh's recent work within an emerging canon of climate fiction. It would thereby forget the continuities that stretch across his literary project more widely and overlook his texts' continued representation of those endangered animals who inhabit the rising water itself. For as we will argue in this chapter, Ghosh's literary engagement with climate in *The Hungry Tide* and *Gun Island* is often inseparable from, and frequently expressed through, his representation of animals, in particular endangered dolphins. To make this argument we will have to become literary dolphin watchers, tracking how Ghosh's texts depict, plot and symbolically rely on the figure of the vulnerable cetacean. Analysing Ghosh's cetacean imaginaries, this chapter therefore sets out to counterbalance the much-discussed "blue" features of Ghosh's writing by interrogating how his fiction negotiates this book's other operative concept, "extinction."

Our approach is necessary because *The Hungry Tide* and *Gun Island* are not just stories of climate change but also texts of the Sixth Extinction—a term that was taken up outside the confines of scientific literature in the years leading up to *The Hungry Tide*'s publication, before then ascending to a wider-reaching if also more vexing cultural prominence in the years since (O'Key 2023). By saying that these are texts *of* the Sixth Extinction we mean first the ostensibly straightforward point that Ghosh sets both novels within a world experiencing what he has his characters call a "dramatic deterioration" in marine life (Ghosh 2005, 266). Into that world he elects a cetologist as his protagonist, and he narrates her efforts to conserve endangered dolphins. Yet narratives are rarely straightforward. One of this chapter's aims is to explore how and to what ends Ghosh leans on these disappearing dolphins, thematically and formally, and to thus consider what this tells us about literature in a time of extinction.

Before this, though, it will be useful to consider why Ghosh centres the figure of the dolphin in these two novels. After all, is it not the whale that has tended to function as the poster species of oceanic degradation? The whale is an icon of climate change, an animal whose figuration, operating firmly "under the sign of extinction" (Huggan 2018, 2), exudes the cultural projections of environmental melancholia with which it is burdened. The dolphin, conversely, has a brighter symbology and has long been conceived—especially in the Global North—as an altruistic do-gooder. With its etymology deriving from the Latin *delphinus*, from the Greek *delphis*, or the womb, the dolphin is imagined as a feminized mother figure, even functioning as a motif of salvation in Christianity (Rauch 2014, 89). This dominant ideological construction of the dolphin was reshaped under the intense geopolitical, technological and cultural conditions of the Cold War. As Mette Bryld and Nina Lykke argue in *Cosmodolphins*, the late twentieth century produced a "complex dolphin iconography" (2000, 183) in which dolphins symbolically mutated into hyper-intelligent cyborgs, gregarious playmates, skilled co-workers and enigmatic New Age aliens whose alterity might unlock new levels of consciousness. Dolphins became central to scientific and military research: from linguistic experiments in human–dolphin communication to the naval conscription of dolphins into mine-detection. And for an urbanized American society yearning to reconnect with nature, Bryld and Lykke write, "the smiling dolphin leaped out as the perfect boundary figure between the extra-terrestrial marine world and that of humankind" (168). In this imaginary the dolphin surfaced as an aesthetically charismatic other and a problem-solving partner. In a word, the Americanized dolphin of the twentieth century stood as an *exceptional* animal whose very exceptionality made it the symbolic equal of society: the "humans of the sea" according to the neurophysiologist John C. Lilly, whose laboratory research and popular writing on bottlenose dolphins consolidated the dolphin's popular status as a uniquely intelligent marine mammal. For Lilly as for others, the "essentially peaceful mission of interspecies communication" (1975, 221) was a redemptive answer to the nuclear arms race. Yet something began to shift with the end of the

Cold War, as the so-called end of history was destabilized, at least in part, by the rise of climate change discourses. The figuration of dolphins' compensatory cheerfulness became overshadowed by the dispirited resignation of whales.

We will have more to say about dolphin iconography below. For now, though, allow us to offer an initial, strong hypothesis: Ghosh's dolphins can be read as postcolonial animals, rejoinders to this culturally dominant Americanized imaginary. They decolonize dolphin aesthetics, focusing not on the iconic and charismatic bottlenose but rather those riverine dolphins who, at least according to one of his characters, are "little floating pigs" that carry no "sex appeal … Or any appeal, for that matter" (Ghosh 2005, 304). We are not the first to point out this contrast in species representation. As Jonathan Steinwand notes, postcolonial novelists like Ghosh "tease and challenge readers to reflect further on our relationship with these charismatic megafauna" (2011, 184) in ways that break with prevailing species logics. However, Steinwand's chapter was written before *The Great Derangement*'s programmatic pronouncements on literature's capabilities, and prior to *Gun Island*'s novelistic negotiation of these capabilities. It is now time, therefore, to reconsider Ghosh's dolphins through a comparative analysis of these two interlinked novels, in which he portrays them as the victims of both American imperialism and its associated industrial development. *The Hungry Tide* states that the Mekong Orcaella "share Cambodia's misfortunes," suffering "the ravages of indiscriminate American carpet bombing" in the 1970s (2005, 305). *Gun Island* imagines a refinery upstream from the Bay of Bengal that thwarts legal challenges to stop it dumping chemical fertilizers into rivers. The refinery—a symbol of how India's twenty-first-century economic development has relied on both environmental degradation and widespread organizational corruption (Bardhan 2022, 7–8)—is responsible for creating "oceanic dead zones" (Ghosh 2020, 104) within the Irrawaddy dolphin's habitat. For Ghosh, the fact that dolphin numbers are falling "below sustainable levels" (305) is inextricable from the histories of imperialist wars and postcolonial nation-building. Ghosh's dolphins thus complicate the politics of dolphin ideology.

And yet Ghosh's figuration of dolphins is by no means static. His cetaceans are complex and contradictory figures. As such, this chapter's key task is to investigate how Ghosh's dolphins metamorphose in shape and shift in function from one novel to the next. Through this we will show that where *The Hungry Tide*'s realism cast dolphins as the objects of scientific inquiry to be watched through binoculars, *Gun Island*'s turn to genre imagines dolphins as mystical bearers of foreknowledge and lively historical agents. We read this shift in representational strategy, from object to subject, from cetological to animist epistemologies, as an attempt by Ghosh to overcome, within the novel form itself, the de-animating forces of a novelistic realism that he believes licenses modernity's destruction of nature. *The Hungry Tide* and *Gun Island* thus plot out two ways of "knowing" nature, as Ghosh's juxtaposition of the knowledge-pursuits of conservation science with a mystical sense of nature's ultimate unknowability suggests a wider reconciliation of modernity with those residual, premodern and altermodern belief systems that modernity dismisses. To track Ghosh's changing representation of dolphins is, then, to track his project's stylistic movement from a mimetic realism towards a more self-consciously magical genre of storytelling. Ghosh's cetacean imaginaries are symptomatic of his wider theoretical attempt to reanimate the novel. This move opens up novel ways of writing ecological relations. But it also risks repackaging, rather than reconstructing, the dominant dolphin iconography, and thus continuing in two differing literary registers the historical tendency to find salvation in the redemptive figure of the dolphin.

Tiger Questions, Dolphin Solutions: Re-reading *The Hungry Tide*

The Hungry Tide is primarily set in the Bay of Bengal, in the shapeshifting and intermittently submerged islands of the Sundarbans, an active delta of streams, mudflats and mangrove forests in which "rivers stray from week to week, and islands are made and unmade in days" (2005, 224). This is an area that muddies the clear distinctions between land, river and ocean, as well as the national borders that separate India and Bangladesh. While Ghosh tells us that the Bengali word "Sundarbans" translates as "beautiful forest," he is at pains to describe this "*bhatir desh* – tide country" (8) as possessing a destructive, sublime beauty. Indeed, the five million people who call this place home, who inhabit and cultivate its many islands and marshy tracts, are at constant risk of sudden tropical cyclones, drawn-out monsoon months and, for crab-catching fishermen in particular, tiger attacks. According to the anthropologist Annu Jalais, with whom Ghosh travelled alongside during his research for the novel, the people of the Sundarbans describe its most southerly islands as being inside "the tiger's lair" (Jalais 2010, 6). "At no moment," Ghosh writes in *The Hungry Tide*'s opening pages, "can human beings have any doubt of the terrain's hostility to their presence" (8). Ghosh portrays the Sundarbans as an ecology that is so stable in its instability that social life is one of constant struggle.

The novel is equally concerned, though, with presenting another side to the tide country. Through a rediscovered notebook—a written testimony by a retired socialist headmaster who worked in the region—Ghosh reveals how the islands are also the site of state-sanctioned injustices performed in the name of species conservation and extinction prevention. Nirmal's notebook, operating as a text-within-the-text, bears fictional witness to a real event: the Marichjhapi massacre of 1979, in which thousands of Bangladeshi refugees found themselves accused of breaking forest preservation laws and trespassing into the "tiger reserve zone" (Jalais 2005, 1760). Under the aegis of the Forest Preservation Act, and in concert with the then World Wildlife Fund, the federal government banned all movement in and out of Marichjhapi. Despite successful legal challenges to this ban, state forces blockaded and later infiltrated the area. This culminated in a mass eviction. Historians have described how military personnel encircled the island, torched the refugees' houses, sank their boats, tear-gassed them, raped and murdered them and dumped their bodies in the river (Mallick 1999, 108–112). *The Hungry Tide* tells this story through Nirmal's first-person recollection of events and Ghosh's own matter-of-fact third-person narration, in which he states that those at Marichjhapi were "forcibly depopulated in order to make room for wildlife conservation projects" (59). *The Hungry Tide* thus characterizes conservation as a conflict over property, resources and dignity. It is a form of "green imperialism" (Guha and Martinez-Alier 1997, 107), a new colonialism after the withdrawal of the old empire, this time in the name of preventing tiger extinction.

Nevertheless, *The Hungry Tide* does not reject conservation tout court. Instead, the novel's plot ultimately calms the conflicts of postcolonial human–animal relations. Ghosh wants to balance the survival of subaltern communities with the clear ecological necessity and moral responsibility of preventing species extinctions. To do this he plots the journey of two protagonists into the Sundarbans, two outsiders who each personify different iterations of postcolonial subjectivity: Piya, a diasporic cetologist who travels to the gulf to study its dolphins; and Kanai, an enterprising translator and self-styled urban cosmopolitan Indian. Through Kanai and Piya's interpersonal and environmental experiences in the Sundarbans, Ghosh stages moments of recognition that lead them towards a sense of responsibility. Kanai commits to re-write from memory his uncle's journal, which is lost in the novel's final set-piece, a climactic storm. Piya similarly realizes that any conservation practice worthy of the name must involve her embedding herself in Lusibari and partnering with the local NGO. Thus, Piya concludes that "I don't want to do the kind of work that places the burden of conservation on those who can least afford it" (397). Ghosh ends the novel by suggesting that his protagonists have undergone their own form of translation: from self-interested individuals into dependable members of the community.

The Hungry Tide has become a touchstone for debates within postcolonial ecocriticism (Weik 2006; Kaur 2007; Huggan and Tiffin 2015; Mukherjee 2010; Heise 2016), echoing the field's ambitions to understand the historically produced conflicts between globalized nature conservation and localized social destitution, and to thereby work towards forms of justice for humans and nonhumans alike. Yet what requires more scrutiny is how dolphins are positioned within this dramatized tussle between human and nonhuman interests. The Sundarbans are home to numerous endangered species: estuarine crocodiles, otters, turtles, cheetals and wild boars, and tigers themselves. Why, then, does Ghosh choose the Irrawaddy dolphin as Piya's object of study? Perhaps because the Irrawaddy, an endangered river dolphin that lives at the hazy border between brackish water and freshwater, neatly represents the peculiarities of the region: its coastline geography of the Sundarbans, in which two different ecotones co-exist, and its cultural confluence, in which the many pujas, punthis and mantras practised by the region's communities blend Hinduism and Islam.

Throughout the novel Ghosh represents dolphins in manifold ways. Chiefly they stand as *Orcaella brevirostris*, the taxonomic objects of Piya's study, a dwindling if not yet critically endangered species who have been commodified by the wildlife trafficking black market (306) and shipped to aquariums across the world. They are depicted as a mysterious species; tracking, surveying and analysing them will be useful for understanding them as a species, but also how best to advocate for them in the public sphere. The dolphins are also represented as victims of wildlife conservation in their own right. When Piya finds the beached body of a newborn dolphin calf with "a huge gash behind the blowhole" (346), killed by a forest conservation patrol boat, Ghosh suggests that the vigilant policing of tiger protection harms island communities and other endangered species alike.

More important, though, is that Ghosh looks to the dolphin in order to stage an encounter between global conservation science and local lifeways. For it is the dolphin that conjoins Piya with Fokir Mandol, a local subsistence fisherman who saves her life at the novel's outset and accompanies her on her river surveys, ultimately shielding Piya from the cyclone at the novel's end, dying in the process. Piya and Fokir are characterized so as to embody seemingly oppositional ideas: the global and the local, cosmopolitan and rooted subjectivities, science and folklore, deep ecology and ecological realism. By contrasting Piya's post-Enlightenment biological sciences with Fokir's traditional knowledge, Ghosh casts dolphins in the shadow of both Linnaean taxonomy and in the illumination of Bengali folklore. Known locally in the Sundarbans as *shushuk*, the dolphins are perceived as the "messengers" (235; 307) of Bon Bibi, a guardian spirit of the forest and ruler of animals whom islanders appeal to for refuge from a hostile nature. This gulf between Piya and Fokir appears unbridgeable when Piya watches on, horrified, as Fokir participates in the communal slaughter of a tiger that had entered a village. Yet when Piya has her fieldwork "epiphany" (123)—that the Irrawaddy engage in a daily cycle of migration between fresh and saltwater, a "novel way of adapting their behaviour to this tidal ecology" (124)—we are invited to think that it pales in comparison to an even greater finding: the "almost miraculous" synthesis of her and Fokir's work, a "seamless intertwining of their pleasure and their purposes" (141). Pleasure and purpose become further interwoven as Ghosh implies a growing romance between them.

By deploying the dolphin as a reconciliatory symbol, however, *The Hungry Tide* obscures the other animal species of the novel: the figure of the tiger. The novel's river dolphins live in proximity to humans; they are trusted co-operators in mutualistic interspecies fishing practices with Sundarbans fishermen, romanticized by Piya as a "remarkable instance of symbiosis between human beings and a population of wild animals" (169). The Bengal tiger, however, is depicted as killing upwards of two hundred people every year, and is associated with historical and continued state violence. While the dolphins are presented as the fishermen's helpers, the tigers are their killers. What is the effect of this narrative move? For critics like Ursula Heise, *The Hungry Tide* compels readers to consider "what might move conservation from tiger scenarios to dolphin scenarios" (2016, 194), from intractability to harmony. Yet for others, Ghosh's novel is taking the easy way out. Piya's ultimate decision to become a rooted cosmopolitan is "only possible," Graham Huggan and Helen Tiffin write, "because the local people have no particular issue with the dolphins; the much more intractable problem of tiger sanctuary is thus displaced by the relatively easy 'dolphin solution', and neither a practical nor a philosophical answer to the situation of the tiger is offered in its place" (2015, 205). In other words, the novel can imagine a fusion of wildlife conservation with community development through the figure of the dolphin, but not the tiger. *The Hungry Tide*'s vision of a globally minded but locally rooted dolphin research programme is clinched by obscuring what Annu Jalais terms the more knotty "people–tiger relation" (8). Hence Sundhya Walther's reading that "the tiger remains so disorderly a presence that the novel is forced to erase it altogether" (2021, 60).

If the tiger remains irreconcilable it is because the novel strives to unsettle the colonial ideology that imagines islanders as living in harmony with nature. Indeed, Ghosh's narration explicitly points to how Piya is disabused of the notion that Fokir is an incomprehensible other with a special and natural access to nature. Yet still, the novel associates Fokir with the figure of the dolphin. And this may well end up reproducing what Archana Prasad calls "ecological romanticism" (2011, 3), the idealizing myth of a naturalized indigenous subjectivity. Indeed, this issue becomes all the more complicated when we consider Bryld and Lykke's observation that the dolphin has long functioned symbolically as something of a "noble savage":

> The dolphin opens the door to a world that promises to redeem our nostalgic longings for a future based in the blissfully innocent lifestyle imagined to have existed prior to techno-civilization. The restored images of the sea as an alluring pastoral source, which in its linkage to womb and breast also incorporates vivid notions of orality, undoubtedly provide an appropriate setting for symbiotic reconnectedness such as the ones displayed by the noble dolphin. (2000, 170–71)

The supposedly innocent lifestyle, the premodern technologies, the oral culture of punthi and folklore—these are what Piya sees when she first looks at Fokir. The novel will work to complicate this blissful portrait, turning romanticization into complexity. Yet it does not extend this gesture to the dolphin, a figure that remains positioned as the symbol of reconnectedness, deployed to circumnavigate, rather than confront, human–tiger conflicts.

Ghosh therefore resolves *The Hungry Tide*'s plot and redeems conservation by turning towards blue extinction and away from green extinction, a move that formally resembles the blue humanities' claim that it offers a methodological and ethical "counterchallenge to the obsessively green visions of ecocriticism" (Mentz 2024, 24). Indeed, we might say that the doctrine of the blue humanities, its motivation and desire, is to shift attention to under-represented forms of life (and death) and thereby transcend "the boundaries and methodologies of land and nation-state-based environmental perspectives" (Perez 2020, 2). But in our view, this "oceanic turn" (Mentz 2009, 2020; DeLoughrey 2017; Oppermann 2023) will be for nothing if it means turning away from tricky terrestrial relations, if it means fetishizing or romanticizing the otherness of the ocean and if it means looking for easy solutions to that which otherwise appears intractable.

Dolphin Ex Machina: *Gun Island* and the Genre Pivot

In the years that followed *The Hungry Tide*'s publication Ghosh began to reflect publicly on why he found it so "extraordinarily difficult" (2016, 33) to write about the nonhuman. An author suddenly inserts a tornado, a flood or a cyclone into their story, as Ghosh does in *The Hungry Tide*. It functions not as description or backdrop but as an object with its own force, an agency that cannot but impact on the story's plotted events. We might be accustomed to think of this as some sort of deus ex machina, or as Ghosh puts it, "a contrivance of last resort" (16). But climate change is changing reality. If extreme weather events with a "very high degree of improbability" (26) become the new ordinary, then Ghosh argues that literary realism will have to find ways of formally incorporating the "improbable" without also shattering the work's reality effect. At stake here is more than just aesthetics. For Ghosh, the entire enterprise of literary realism is built on the suppression of the nonhuman—both a cause and effect of a wider historical conflict in modernity between dominant rationality and residual superstitions. He believes that the planet is dying because Western modernity imagines nature as inert matter that can be materially exploited without consequence. For him, the task of literature in the twenty-first century is to shatter modernity's "dream of silencing the nonhuman" (65), to reveal the ways in which this myth, as myth, can never be fully realized. As Ghosh writes in his more recent book of essays, *The Nutmeg's Curse*, the "erasure of nonhuman voices from 'serious' literature" has helped create "that blindness to other beings that is so marked a feature of official modernity. It follows, then, that if those nonhuman voices are to be restored to their proper place, then it must be, in the first instance, through the medium of stories" (2021, 204).

It is tempting to read *Gun Island* as a direct response to this form-problem, as what Ursula Kluwick calls a "literary reply to his own theoretical deliberations" (2020, 8), in which the novel's stylistic decisions are read as confirmations of the author's preoccupation with climate. Most notably, *Gun Island* breaks with the third-person narration of *The Hungry Tide*. Switching to a first-person prose enables Ghosh to narrate the nonhuman from a partial rather than universalizing point of view, framing the story from the vantage point of Dinanath Dutta, or Deen, an antiquarian bookseller in his sixties who lives, like Ghosh himself, in Brooklyn. Kanai, the self-styled urbane businessman of *The Hungry Tide*, is here relegated to a minor character, a "distant relative" (2020, 4) of Deen's who functions as little more than a messenger, a character-node whose opening encounter with the protagonist kick-starts the plot. Deen, an erstwhile reader of Bengali folklore, is tasked with visiting the shrine of the Gun Merchant, or Bonduki Sadagar, a legend of the Sundarbans and subject of a Bangla verse epic. What begins for Deen as a straightforward question of philology quickly mutates into a quasi-mythic quest with planetary stakes. Where Deen chases the Gun Merchant story around the world, another character—Piya, returning from *The Hungry Tide*—continually chases pods of endangered dolphins.

Gun Island takes up a different narrative perspective to *The Hungry Tide*, but the novel still shares a similar narrative arc: Deen, who occupies the role that Kanai once filled as the sceptic, finds himself having "a kind of awakening" (237). He is continually amazed, astonished, struck and left speechless—the words pile up across the novel, with variants on "surprised" occurring almost fifty times—by coincidences that are felt locally but always interpreted as manifestations of an ambiguous global network, "a world of impersonal systems" (236) and "improbable intersections between the past and present" (201) that includes the premodern spice trade, forest fires in California and rarely visited shrines in the Sundarbans. Deen ends the novel in Venice, having traced the Gun Merchant story to the city's former Jewish ghetto. There he meets groups of Indian and Bangladeshi migrant labourers, and notes that—at least from the sky—it was "possible to mistake the Venetian lagoon for the Sundarbans" (162). Blurring the boundary between the Global North and South also means scrambling historical time: here the present recalls the past, and the contemporary world of climate migrations overlaps with the seventeenth-century world that once produced the Bonduki Sadagar story.

Reclaiming literature's ability to move from one "exceptional event to another" (2016, 16), as Ghosh writes in *The Great Derangement* of pre-novelistic storytelling, *Gun Island* attempts to derange realism through the catastrophic and fundamentally weird new reality of a climate changed and changing world. For this is a more geographically global, politically world-systemic and aesthetically planetary novel than *The Hungry Tide*. Through a globe-spanning story, Ghosh re-scales his thematic focus in order to register the transnational reach of climate change, and to evoke a deep sense of connection, at times an internationalist solidarity, between the North and South. Furthermore, where *The Hungry Tide* adheres to a recognizable realism of the historical novel as its baseline register of expression, *Gun Island* wholeheartedly embraces genre forms like the adventure, mystery, crime, thriller, epic and myth—the very forms that Ghosh characterized in *The Great Derangement* as valences of the narrative imagination that modernity wishes to banish. Crucially, by having Deen strike up a romantic relationship with Piya, Ghosh reconstructs and realizes the romance plot that *The Hungry Tide* once suggested and then thwarted.

The figure of the dolphin is key to Ghosh's invention of a global and genre-multiplying novel. Indeed, *Gun Island*'s turn towards dolphins continues *The Hungry Tide*'s turn away from tigers. Tigers are almost entirely absent here, described only in monstrous terms as having "infested" the Sundarbans (8) and inspired "hideous" (57) depictions. For a reader familiar with *The Hungry Tide* their absence haunts the new novel—a further displacement of the tiger question. By continuing Piya's story *Gun Island* maintains a focus on dolphins, again contrasting the secular explanatory force of marine biology with folk stories of nature. Piya enters the novel with her record books and GPS tracker, having now, in the space between *The Hungry Tide*'s past and *Gun Island*'s present, closely studied a single pod of dolphins, mapping their "movements over many years, following their diurnal, seasonal and annual migrations" (100). Piya's research has shown that sea level rises have increased the salinity of West Bengal's coastal waterways, forcing the pod further upriver, escalating their chances of being caught in fishing nets or struck by riverboats.

Piya has spent considerable time studying the matriarch and oldest living member of the pod, officially designated as RN_1 but informally known as Rani—named so by Tipu, a digitally literate and culturally Americanized Indian teenager who is the son of *The Hungry Tide*'s Fokir. The dolphin's double name signals the ways that *Gun Island* juxtaposes a strict scientism with the animating force of literary characterization. Piya recounts to Deen that although she had been studying Rani for a number of years, it was only after she had rescued her from a stretch of nylon netting that the dolphin "had begun to make eye contact ... in a manner quite different from other members of the pod—a manner that suggested something more than mere recognition" (101). Rani, then, becomes for a moment a legitimately minor character, and Ghosh deliberately bends his syntax to emphasize Rani's animacy: the pronoun "it" becomes a personified "she": "More than any other members of the pod it was she who had helped Piya track the family's migrations" (101). Yet under the worsening conditions of sea level rise and river pollution, the pod's migrations have become irregular and unpredictable. Piya receives a concerning alert on her GPS and, travelling out into the tide country, discovers that the members of the pod have "beached themselves, all at the same time" (107). Later in the novel dozens of Irrawaddy dolphins are stranded on the island of Garjontola, a mass beaching that implies functional extinction. "I don't know what to tell you, Deen," Piya says. "It was the most devastating thing I've ever seen. So many of them, throwing themselves up on the shore. I've heard of other cetaceans doing this but never *Orcaella*" (195).

In *Gun Island*, then, registrations of species extinction are coupled with assertions of nonhuman animacy. Strandings do not just *happen to* dolphins (107), the novel suggests, but are suicidal actions taken by dolphin pods in order to warn of global climatic derangement. This notion of an animate nature reaches its climax in the novel's own climactic scene, when a boat carrying migrants and refugees from Ethiopia, Eritrea, Somalia and Bengal arrives at the Italian coast. Refused entry to numerous ports (a nod, presumably, to the standoff in 2018 between the right-wing Italian government and Fortress Europe over the MS *Aquarius*), the boat has become "a symbol": for the left, "of everything that's going wrong with the world—inequality, climate change, capitalism, corruption, the arms trade, the oil industry" (218); for the right, "the overturning of a centuries-old project that had been essential to the shaping of Europe" (304–305).

Ghosh has nature suddenly intervene into this contested present. Various marine species—long-finned pilot whales, fin whales, sperm whales, Risso's dolphins, striped dolphins—congregate alongside the boat, appearing to guide the migrants towards the safety of the harbour. An Ethiopian woman, read as an incarnation of the snake goddess Manasa Devi from the Gun Merchant story, stands with her arms aloft, "a halo of birds spinning above her, while down in the water a chakra of dolphins and whales whirled around the boat" (307), a storm that sees "the creatures of the sky and sea rising up" (306).

In our view, this plot device and narrative development resembles what we could call a *dolphin ex machina*, a "miraculous" (293) denouement of cetacean animacy in which the nonhuman world rushes to the side of the stranded human. Indeed, the Mediterranean Sea—the final setting of Ghosh's novel—is steeped in dolphin mythology: in the story of the Oracle of Delphi, which echoes the Gun Merchant myth, Apollo is said to have killed a snake guarding the site at Mount Parnassus, before then metamorphosing into a dolphin in order to guide priests towards the new temple. By recalling this mythology Ghosh simultaneously deepens and broadens his text: deepens because he shows how human–animal relations extend back across history; broadens because he is connecting the Bonduki Sadagar story to others from around the world. This then provides the context for the book's final message: "*Unde Origo Inde Salus*—from the beginning salvation comes … that the possibility of our deliverance lies not in the future but in the past, in a mystery beyond memory" (312). *Gun Island* ends, therefore, with the idea that only the past can redeem a damaged present. This is not a yearning for the glittering past of the Golden Age, even if it has its own nostalgic logic: a call to turn to all the temporalities and forms of life that modernity consigns to the past, to backwardness, to forgetting. It is, rather, a community-making ending, one that conjures out of its mythic realism forms of multicultural and multispecies togetherness: the migrant and the animal, side-by-side.

Conclusion: Nature Is Healing

Soon after *Gun Island* was published the world was ravaged by the Covid-19 pandemic. Many states instituted population-wide lockdown measures to halt the virus' spread. Within a month of these lockdowns, a number of social media posts started circulating that proudly announced that nature was healing. The most iconic of these misanthropic and patently fallacious posts focused on Venice, proclaiming that the city "hasn't seen clear canal water in a very long time. Dolphins showing up too." In the sudden absence of human traffic, "Nature just hit the reset button on us" (Bale 2020; Allison 2020, 4.2). These tweets, written before *Gun Island* had even received its paperback printing, can be read in light of the novel's own intentional warping of chronology and causality. For is this image of nature hitting the reset button, this faith in water as a rapidly self-renewing force, not strikingly similar to the way that *Gun Island* ultimately imagines the Mediterranean?

By closely reading depictions of dolphins, this essay has drawn out how Ghosh engages the harsh realities of postcolonial planetary ecology through animal figures. Indeed, in both *The Hungry Tide* and *Gun Island* Ghosh seeks to find narrative ways of freeing nature from its disenchantment. His novels juxtapose different ways of looking at dolphins; the scientific and the magical, the global and the local. As such, dolphins are portrayed as highly charismatic marine creatures that behave in line with observable and recordable ecological dynamics and patterns. Yet they are also presented as unknowable others that elude surveillance, as figures associated with the vastness and enigma of the ocean itself, representing an agency beyond human comprehension. Their figuration sheds light on the localized impacts of habitat destruction while also generating a kind of environmental ethics by provincializing knowledge itself: only by acknowledging the limits of our understanding, these novels seem to say, can we cultivate a respect for the lively autonomy of dolphins. In *The Hungry Tide*, the vastness of the ocean and its creatures spark epistemic and historical questions that decolonize biological sciences. This idea is pushed further in *Gun Island*'s deliberate attribution of agency to animals. Nevertheless, neither text abandons the sciences—Piya's return as a character confirms as much. Instead, these novels aim for a vision of biology and indigeneity as complementary forms of human–animal relations.

However, as our close readings have shown, these novels are both literary interventions into the Sixth Extinction and products of this cultural moment. Their turn towards endangered animals, and in particular the ways they look at the disappearing dolphins, are part of a symbolic quest for a redemptive otherness that can solve the extinction crisis. Whether writing in the overt realism of *The Hungry Tide* or the more animistic register of *Gun Island*, Ghosh still ends up deploying dolphins for the same purpose: as rescuers that resolve the given novel's plot. From the dolphin solution of the earlier novel to the *dolphin ex machina* of the latter, these novels replay the mythic figuration of the dolphin as a companionable saviour of a doomed humankind.

In *The Hungry Tide* and *Gun Island* blue extinction therefore becomes the occasion for plotted resolution, ecological reanimation and literary-cultural re-enchantment. Dolphins operate here as both victims of modernity and its sunny saviours. Ghosh's project thus offers critical representations of human–animal relations in a time of biodiversity loss, but there remains a compensatory nostalgia that animates these figurations of nature. What all of this tells us is that Ghosh's environmental turn, by bestowing nature with a plot-changing agency, also risks romanticizing it. At stake in our reading is a view that, in dialogue with one another, animal studies and postcolonial studies offer the blue humanities an exacting attention to mythic ideas of nature, and a ruthless criticism of literature and criticism's symbolic instrumentalization of animals for redemptive ends. Dolphins might well redeem us. But who will free them?

References

Allison, Marcia. 2020. 'So Long, and Thanks for All the Fish!': Urban Dolphins as Ecofascist Fake News During COVID-19. *Journal of Environmental Media* 1 (1): 4.1–4.8.

Anand, Divya. 2008. Words on Water: Nature and Agency in Amitav Ghosh's *The Hungry Tide. Concentric* 34 (1): 21–44.

Bale, Rachael. 2020. Why Do People Want So Badly to Believe This Fake Story Is True? *National Geographic*, 27 March. https://www.nationalgeographic.com/animals/2020/03/why-do-people-want-so-badly-to-believe-this-fake-story-is-true/. Accessed 2 April 2024.

Bardhan, Pranab. 2022. The 'New' India: A Political-Economic Diagnosis. *New Left Review* 136 (3): 5–27.

Bryld, Mette, and Nina Lykke. 2000. *Cosmodolphins: Feminist Cultural Studies of Technology, Animals and the Sacred*. London: Zed Books.

DeLoughrey, Elizabeth. 2017. Submarine Futures of the Anthropocene. *Comparative Literature* 69 (1): 32–44.

Ganguly, Debjani. 2020. Catastrophic Form and Planetary Realism. *New Literary History* 51 (2): 419–453.

Ghosh, Amitav. 2005. *The Hungry Tide*. London: HarperCollins.

———. 2016. *The Great Derangement: Climate Change and the Unthinkable*. London: Chicago University Press.

———. 2020. *Gun Island*. London: John Murray.

———. 2021. *The Nutmeg's Curse: Parables for a Planet in Crisis.* London: University of Chicago Press.

Guha, Ramachandra, and Martinez-Alier. 1997. *Varieties of Environmentalism: Essays North and South.* London: Routledge.

Heise, Ursula K. 2016. *Imagining Extinction: The Cultural Lives of Endangered Species.* London: University of Chicago Press.

Huggan, Graham. 2018. *Colonialism, Culture, Whales: The Cetacean Quartet.* London: Bloomsbury.

Huggan, Graham, and Helen Tiffin. 2015. *Postcolonial Ecocriticism: Literature, Animals, Environment.* 2nd ed. London: Routledge.

Jalais, Annu. 2005. Dwelling on Morichjhanpi: When Tigers Became "Citizens," Refugees "Tiger-Food". *Economic and Political Weekly* 40 (17): 1757–1762.

———. 2010. *Forest of Tigers: People, Politics and Environment in the Sundarbans.* London: Routledge.

Kaur, Rajender. 2007. "Home Is Where the Oracella Are": Toward a New Paradigm of Transcultural Ecocritical Engagement in Amitav Ghosh's *The Hungry Tide. Interdisciplinary Studies in Literature and Environment* 14 (1): 125–141.

Kluwick, Ursula. 2020. The Global Deluge: Floods, Diluvian Imagery, and Aquatic Language in Amitav Ghosh's *The Hungry Tide* and *Gun Island. Green Letters* 24 (1): 64–78.

Lilly, John C. 1975. *Lilly on Dolphins: Humans of the Sea.* New York: Anchor Books.

Mallick, Ross. 1999. Refugee Settlement in Forest Reserves: West Bengal Policy Reversal and the Marichjhapi Massacre. *The Journal of Asian Studies* 58 (1): 104–125.

Mentz, Steve. 2009. Toward a Blue Cultural Studies: The Sea, Maritime Culture, and Early Modern English Literature. *Literature Compass* 6 (5): 997–1013.

———. 2020. *Ocean.* New York: Bloomsbury.

———. 2024. *An Introduction to the Blue Humanities.* New York: Routledge.

Mukherjee, Upamanyu Pablo. 2010. *Postcolonial Environments: Nature, Culture and the Contemporary Indian Novel in English.* Basingstoke: Palgrave.

Nuttall, Sarah. 2020. Pluvial Time/Wet Form. *New Literary History* 51 (2): 455–472.

O'Key, Dominic. 2023. Extinction in Public: Thinking through the Sixth Mass Extinction, Environmental Humanities, and Extinction Studies. *Environmental Humanities* 15 (1): 168–186.

Oppermann, Serpil. 2023. *Blue Humanities: Stories Waterscapes in the Anthropocene.* Cambridge: Cambridge University Press.

Perez, Craig Santos. 2020. 'The Ocean in Us': Navigating the Blue Humanities and Diasporic Chamoru Poetry. *Humanities* 9 (66): 1–11.

Prasad, Archana. 2011. *Against Ecological Romanticism: Verrier Elwin and the Making of an Anti-Modern Tribal Identity.* New Delhi: Three Essays Collective.

Rauch, Alan. 2014. *Dolphin*. London: Reaktion.

Sankaran, Chitra. 2012. Introduction. In *History, Narrative, and Testimony in Amitav Ghosh's Fiction*, ed. Chitra Sankaran, xiii–xxxii. Albany: SUNY Press.

Steinwand, Jonathan. 2011. What the Whales Would Tell Us: Cetacean Communication in Novels by Witi Ihimaera, Linda Hogan, Zakes Mda, and Amitav Ghosh. In *Postcolonial Ecologies: Literatures of the Environment*, ed. E. DeLoughrey and G.B. Handley, 182–199. Oxford: Oxford University Press.

Walther, Sundhya. 2021. *Multispecies Modernity: Disorderly Life in Postcolonial Literature*. Waterloo, ON: Wilfred Laurier University Press.

Weik, Alexa. 2006. The Home, the Tide, and the World: Eco-cosmopolitan Encounters in Amitav Ghosh's *The Hungry Tide*. *Journal of Commonwealth and Postcolonial Studies* 13 (2): 120–141.

Wilton, Demi. 2021. "We are the Dispossessed": Displacement, Knowledge Production and Bare Life in West Bengali Climate Fiction. *Parallax* 27 (3): 344–361.

CHAPTER 3

When the Whales and Seals Are Gone: Verne, Toussenel, and the Brave New Ocean

Killian Quigley

In 2010, the oceanographer Jeremy Jackson delivered an energetic and alarming public lecture at the Smithsonian Institution's National Museum of Natural History in Washington, DC. Entitled "Brave New Ocean," the talk pitched Jackson's widely cited prognostications of mass marine extinction for a lay audience.[1] It invoked a heterogeneous array of cultural references along the way, from Aldous Huxley's dystopian novel *Brave New World* (1932) to Roger Corman's cult black comedy *The Little Shop of Horrors* (1960) to Rachel Carson's classic anti-pastoral nonfiction *Silent Spring* (1962). The presentation also made fleeting mention of *Twenty Thousand Leagues Under the Seas* (*Vingt mille lieues sous les mers*, 1869–1870), Jules Verne's phenomenally and enduringly popular submarine adventure. Jackson's slides briefly highlighted a memorably odd passage from Verne's novel in which its narrator, the naturalist Pierre Aronnax,

[1] For an exemplary instance of those prognostications see Jackson (2008).

K. Quigley (✉)
Australian Catholic University, Melbourne, VIC, Australia
e-mail: Killian.Quigley@acu.edu.au

© The Author(s), under exclusive license to Springer Nature Switzerland AG 2024

V. Fibisan, R. Murray (eds.), *Blue Extinction in Literature, Art, and Culture*, Palgrave Studies in Animals and Literature,
https://doi.org/10.1007/978-3-031-69910-8_3

foretells a time when overhunting has caused whales and seals to disappear altogether from the ocean. Those future seas, muses Aronnax, will have become "cluttered with octopi, medusae, and squids." Worse yet, their waters will have turned into vast "hotbeds of infection in the absence of those huge mammals whose stomachs were intended by God to keep the oceans clean" (Jackson 2010).[2]

The present chapter departs from Jackson's glancing invocation of *Twenty Thousand Leagues Under the Seas* (*20TL*) to recover the pasts and futures of Aronnax's worry. The source of his fantasy of uncleanliness, named openly by Verne, is the writing of the French naturalist Alphonse Toussenel, whose *L'Esprit des bêtes* (1847, 1853–5) registered vivid concern for the long-term consequences of whaling. In *20TL*, Toussenel's grim prospect is just one aspect of a broadly cetaceous mood, which emerges in the novel's early sections with the pursuit of a pseudo-whale and is sustained through the high drama of a late visit to the "basin of the South Pole," where animals "too fiercely hunted by mankind" travel to seek "refuge" (Verne 2019, 284). The narrative's formative relationship to marine locomotion is, in these and other regards, constitutively whaly. "Good" cetaceans and those who would "destroy" them are contrasting features, meanwhile, of a moral worldview espoused by Verne's iconic antihero Captain Nemo, creator of the extraordinary submersible *Nautilus* (2019, 262–3). Nemo's moral system intersects, in turn, with his sense of the ocean's "huge role" in "the globe's overall economy," a role that Aronnax's ghastly prevision tacitly affirms (Verne 2019, 110). To read *20TL* for the whales is to recognize the mingled novelistic powers of scenes of cetacean destruction and prophecies of marine impurity, infectiousness, and collapse.

Commentators have made little of the Verne-Toussenel connection, but *20TL*'s apparent prescience regarding the precarious health of oceanic ecosystems has been repeatedly attested. Three decades ago, the economist Steve DeCanio described Aronnax's forecast as "one of the first articulated theories of ecocatastrophe" (1994, 85). More recently, the ocean ecologist Enric Sala cited the passage as evidence both of Verne's "visionary" achievement and of Nemo's status as "the first marine conservationist

[2] I draw this paragraph's quotations directly from Jackson's slides; he has consulted Mendor T. Brunetti's translation of *20TL*. This essay relies on William Butcher's rendition (1998, updated 2019), the most complete and scrupulous extant.

in history" (2015, 108).[3] The idea that *20TL*—which *Scientific American* once called "the world's first novel of oceanography" (Evans and Miller 1997, 95)—might stand at the beginning of some modern disposition towards apprehending ecology and extinction at sea suggests that its relationship with its own materials might be worth comprehending more thoroughly, not least when those materials are freighted with generalized misanthropy and specific prejudice. In the following pages, I aim to make a little contribution to such comprehension while also contemplating some issues that I take to be of more general pertinence for this book's topic. Those include how to understand a work like *20TL* as simultaneously a discrete literary text and an artefact of extra-literary citation—and how to locate the place *of* the literary amidst the epistemological and communicative projects of marine science. That place would seem to be simultaneously fundamental and fraught, and better characterizing it may serve some purpose for all of us engaged with the figuration of future seas.

Hunting, "Destroying," and Creaturely Virtue

20TL is first of all the story of a hunt. The year is 1866, and mariners' reports are piling up of the existence at sea of an "object or creature" of "extraordinary speed," "surprising agility," and "peculiar vitality" (Verne 2019, 5). If this is a whale, it is one built to previously unimaginable proportions and equipped with a correspondingly powerful weapon: some giant narwhal, heretofore unknown? That piece of speculation comes from Aronnax, whose special expertise—he is the author of *The Mysteries of the Ocean Deeps*, in two volumes—earns him an invitation to cast off with the *Abraham Lincoln*, a frigate-cum-whaler outfitted "with everything needed to catch the enormous cetacean." The vessel's unique accoutrements include the latest American cannon as well as "the king of the harpooners," an emotionally volatile Canadian named Ned Land (Verne 2019, 21). Land eventually joins Aronnax and his "manservant," the phlegmatic Conseil, as Nemo's unintended guests after a violent encounter with the real "monster"—a "species of submarine boat"—leaves them cast overboard and the *Abraham Lincoln* crippled (Verne 2019, 16, 38, 42). The giant narwhal proves to be a chimera, but throughout what

[3] Sala would seem to put Aronnax's words in Nemo's mouth, an error which may inadvertently signal the fact that where matters of marine science are concerned, the naturalist and the captain frequently speak with one voice.

follows, the *Nautilus* stays whalish. It moves "with the skill of a cetacean" and attracts the persistent attention of hunters who mistake it "for an enormous whale of great value" (Verne 2019, 241, 252).

A "whale-seeming submarine" was the science fiction and fantasy writer Ray Bradbury's phrase for a craft so surpassingly inventive as to swim comfortably among marine mammals, and to outrage humankind (1962, 3). Nemo, meanwhile, is outraged *by* humankind for what he feels are its tendencies towards cruelty and despotism. Such literally superficial phenomena—they have no place "30 feet below"—are embodied above all by the tyrants whose terrestrial rule Nemo conspires vaguely to resist (Verne 2019, 63). But at sea, and in the incidents that comprise Verne's novel, humanity's worst aspects are typified by the practices of whaling and the persons of whalers. In the Southern Ocean, the setting of what are arguably the best-wrought episodes in the *Nautilus*'s circumnavigation of the globe, Nemo informs Aronnax of the "destructive rage" of "British and American whalers" who "indiscriminately massacre" seals, walruses, and cetaceans, including "pregnant females and their young." The walruses, Nemo explains, "will soon be destroyed down to the last." Of the seal-herds of the South Shetland and South Orkney Islands, he reflects that the hunters have left "behind them, in place of vivacious life, the silence of death." "Hunting simply to destroy" is the crime Nemo lays at the feet of Land and his "fellows," whose actions the captain construes as deplorably unattached to any scheme that might reasonably justify their consequences (Verne 2019, 268, 282).

It bears emphasizing that what is at issue here is not hunting in general. As Verne's editors have sometimes been at pains to explain, *20TL* is chock-full, jarringly, of unlamented creaturely death.[4] More often than not, such death is only startling by virtue of how it becomes forgotten, instantly and flagrantly, by a narrative that is usually busy surging onwards to some other incident. This is the case even when a hunt is accompanied by prospects of extinction and exhaustion. During a diving excursion in the Pacific Ocean, Nemo shoots a sea otter (*Enhydra lutris*) whose "precious" kind has been so far "tracked and hunted down by fishermen" as to be destined, "probably," to "soon become extinct" (Verne 2019, 105). Floating at the surface of the Red Sea in view of a rare dugong, Conseil wonders whether "if by chance this were the last of its race, [it would] not be better

[4] Citing "Verne's own opposition to hunting," Butcher believes it "possible" that even the hunting events that pass uncondemned are locations of critique (Verne 2019, 385).

to spare it—in the interests of science?" (Verne 2019, 201) Like Nemo's otter, Conseil's dugong does not survive the scene. Episodes like these are characterized by a disjunction between exposition and action that is sometimes low-key but is as often arrestingly blatant.

Explaining this structuring ambivalence may require nothing more than a forthright acknowledgement of *20TL*'s genre. For the OuLiPo experimentalist Georges Perec, Verne's example proved the liberatory potential of "the archetypes of the adventure story" for the literary imagination (Bellos 1999, 431). Adventure's conventions, one might plausibly say, simply demand that no opportunity for activity—perhaps especially sanguinary activity—goes unexploited. And this interpretive line would seem to complicate any attempt to read *20TL*, as does the novelist Margaret Drabble, for "early signs of awareness of ecology and environment and entropy" (2014, 51). But as later sections of this chapter will consider in more detail, another sort of Vernian ecocriticism is possible, one that apprehends the awkwardness of the otter and dugong passages as the signs not of an ecological sensibility raring to break free from the trappings of cultural form but of the perplexities that continue to define so much of the relationship between extinction in theory and human behaviour in fact.

If *20TL* does offer anything like an internally consistent account of the ethics of killing, it is one situated at the nexus of a critique of intemperate, indiscriminate, and annihilative hunting and a complex scheme of creaturely virtue. From this juncture, which is never entirely stable, emerges a paramount concern for the fates of whales and the foibles of whalers. In one emblematic exchange, Nemo condemns Land's occupation for "destroying the Antarctic whale like the right whale."[5] As previous quotations have indicated, the verb "destroy" (*détruire*) is invoked expressly, here, for the specific sense it conveys of complete undoing—of not just injuring a species but extinguishing it. Aronnax receives Nemo's words as a "lesson" that is not only "moral" but "right": "the barbaric and unthinking relentlessness of the hunters," he realizes, "will one day eliminate the last whale from the ocean." We have seen, of course, that the idea of extinction frequently fails to elicit pointed or sustained distress from Verne's characters. What appears to make the difference, in this case, is Nemo's description of the "Antarctic" and right whales as not only threatened animals but "inoffensive and good creatures" (Verne 2019, 263).

[5] "Antarctic whale" is Butcher's translation of Verne's *baleine australe*, a name for the southern right whale (*Eubalaena australis*).

In whaling, *20TL*'s animal-killing acquires both an unusual degree of extirpative potential and an exceptionally admirable—and therefore gallingly problematic—object. Contemplating the spectacle of abundant cetaceans in the Southern Ocean, Aronnax muses that "these animals, under outrageous pressure from hunting, have taken refuge in the higher latitudes" (Verne 2019, 259). With passages like this one, the novel's critique of extinctive whaling and admiration for (certain) whales coincide with its sympathy for such parts of the globe as ostensibly lie at or past the frontiers of human spoliation. Nearing the South Pole, Aronnax records right whales, humpbacks, and finbacks, "playing in schools in the peaceful waters." The observation is made at the surface—Aronnax also notes some "jellyfish of great size rocking on the waves"—but in their tranquillity these "waters" emblematize an oceanic character that is otherwise available only at depth (Verne 2019, 284). "Perfect peace abides," for Nemo, "in the heart of the sea," a refuge for "independence" and freedom as well as biological life—and *from* the "power," "influence," and "authority" of "men" (Verne 2019, 62–3). What the captain finds "damnable" in the whalers therefore derives, not least of all, from their signifying an unusually intense affront to the arguably romantic desire for some parts of earth's waters—if not its lands—to remain beyond humanity's ken (Verne 2019, 263).

This affront takes creaturely form, in Verne's southern polar episodes, through a gruesome encounter with—of all things—sperm whales. A group of those animals are prevented from harassing some "baleen whales" when the *Nautilus* wreaks a shockingly brutal attack upon them. Sensing a rare chance to seize the moral high ground, Land disparages the act as "just butcher's work." Not so, retorts Nemo: "it was a massacre of evil animals" (Verne 2019, 265). It is as though in *20TL*'s ambiguous ethics, the cachalots come to embody the worst aspects of the same human profession that had become renowned for pursuing them—and their bodies are made to bear the brunt of the *Nautilus*'s righteous vengefulness. For its part, the whale-seeming submarine becomes a weapon of resistance in a hunt of baleen by toothed cetaceans, as well as in an immeasurably wider conflict between Nemo's misanthropic submarine revolutionism and the forces—human and otherwise—of offense. How those uncoincidentally modern energies intersect with the novel's sense of oceanic "economy" is the subject to which this chapter now turns.

Cleanliness, Disease, and the Economy of the Ocean

"Look at this ocean," demands Nemo of Aronnax during one of their many dialogues. "Is it not endowed with an authentic life of its own?" (Verne 2019, 109). Nemo understands the sea as a kind of organism, a view he credits explicitly to the American oceanographer and naval officer Matthew Fontaine Maury, author most famously of *The Physical Geography of the Sea* (1855). Maury had figured himself (and his reader) as "surrounded" by a "beautiful system of cosmical arrangements and terrestrial adaptations," a system instantiated in part through a distinct "economy of the ocean" (1844, 181). The latter phrase partakes of an older and wider tradition of comprehending life on earth in terms of an "economy of nature"—a multifarious tradition that Donald Worster explored as a kind of forerunner to modern "ecological thought" (1977, viii–ix). For Nemo, the most pivotal aspect of the economy of the ocean is its "genuine circulation"—its having a "pulse and arteries," even "contractions," just "like the blood in animals." This divinely ordained state of affairs operates through "currents and counter-currents" and affiliated mechanisms of "interchange": between "tropical and polar waters," to be sure, but also between the surface and "the depths." Meantime the sea's salinity serves as a "stabilizer of the globe's overall economy" while its "billions of animalculae" construct "the limestone landmasses" and facilitate the generation of "continuous life": "life more intense than on land, more exuberant, more immeasurable, blossoming in every part of the ocean" (Verne 2019, 109–10).

If Nemo's own exuberance testifies to the wondrous productions of the sea's economy, it also signifies the stakes involved in its circulation's good functioning. Late in *20TL*, Aronnax contemplates the "oceanic river" called the Gulf Stream, discovery of the "true source" whereof he accords to "Commander Maury." In its "role as moderator," that system of currents is "charged with re-establishing the equilibrium" between "the tropics" and "the northern waters." "We have to hope," broods Aronnax, that the Gulf Stream's "regularity is maintained; for if its speed and direction were ever to change, as some believe is happening already, the climate of Europe will undergo changes of an unforeseeable scope" (Verne 2019, 319–20). For Verne's naturalists, the marvels of cosmical arrangement are never unhaunted by the terrible possibility that their economy might someday collapse. Currents, wrote Maury, "not only dispense heat, and moisture, and temper climates, but they prevent stagnation in the sea; and by their active circulation, transport food and sustenance for its inhabitants from one region to another, and people all parts of it with life and animation" (1844, 181). The possibility of this activity going awry looms in the background of a novel that is animated, throughout, by the future's uncertain prospects.

Of all the "active and important parts" assigned to currents in Maury's scheme, their guarding against marine "stagnation" may be the least familiar to twenty-first-century readers (1844, 181). A body's lack of motion carried vivid associations, for nineteenth-century science and its antecedents, with "putrefaction, corruption, and decay"—and frequently evoked the corollary belief that diseases find hospitable the absence of elemental circulation (Lawrence 1996, 82). For *20TL*, the idea of a stagnant sea offers the startling notion of pestilential rot on a grand scale. Having encountered some manatees off the coast of Suriname, Aronnax reflects upon the "important role" that "far-seeing nature" has given them: "like the seals, they graze on the submarine prairies and thus destroy the accumulated grass which obstructs the mouth of tropical rivers." But the destroyers of grass were themselves "almost entirely destroyed" by "men," and in consequence "decaying grass poisoned the air, and the contaminated air produced yellow fever which is destroying these admirable countries." Overhunting set in motion a chain of effects which has culminated, explains Aronnax, in "poisonous vegetation" having "proliferated in the warm seas … from the mouth of the Rio de la Plata to Florida!" (Verne 2019, 308).

However credible it may be, Aronnax's theory is remarkable for the way it charts a causal relationship between the near-extinction of a marine animal and an intercontinental economic—and epidemiological—catastrophe. And it is with regard to this relationship that Verne's debt to his compatriot Toussenel becomes unmistakably apparent. In *L'Esprit des bêtes*, manatees and seals are recognized as some of humankind's greatest assistants (*auxiliaires*), tasked with consuming the submarine grasses that traverse the great river mouths of tropical America: the Orinoco, the Amazon, and so forth. So long as their work of divine hygiene (*sa fonction hygiénique providentielle*) carried on, the Gulf of Mexico stayed free from the plague of yellow fever, which is supposed to emanate from marine vegetation as it putrefies. These circumstances were irrevocably altered under the influence of European trade, which demanded that the manatees be destroyed for their oil. Such forces of commerce—which are effectively personified by Toussenel's prose—thus cleared the way for the "*vomito* [*sic*] *negro*" to enlarge its reach. The disease came to reign, eventually, from the mouths of the Plata to the Mississippi, from Rio de Janeiro to New Orleans—and its occasional appearances in Europe only presage its fuller establishment there (Toussenel 1868, 154–5).

Yellow fever, which is caused by a mosquito-borne virus related to dengue and West Nile, is thought to have been introduced to the Americas via the transatlantic slave trade (Bruijn 2009, 52). Its operations there were objects of appalled fascination for mid-nineteenth-century European media publics and sustained interest for Verne, whose *From the Earth to the Moon* (*De la terre à la lune*, 1865) had also invoked the disease (1978, 62). For Toussenel, the *vómito negro* furnished a potent illustration not only of what Maury called "cosmical arrangements" but of a *moral* economy of European and Euro-colonial society. The killing of manatees has already been mentioned. Comparably poor in virtue are the bourgeoisies of New York, London, and Amsterdam, whose commitments to luxury are purported to be incommensurable with any serious regard for the interests either of mankind in general or of the natural relations which might ordinarily serve those interests. Pondering whether the Gulf of Mexico's manatees could be replaced by marine turtles, Toussenel scoffs at the idea that the gourmands would give up their soup. After all, have the awful connotations of imperial agriculture in the Caribbean ever compelled anyone to stop adding sugar—*sang de nègre*—to their coffee? Toussenel thinks not (1868, 155).

For *L'Esprit des bêtes* as for *20TL*, the consequences of manatee and sea-turtle annihilation ultimately pale in comparison with what the disappearances of two other kinds of aquatic animal hold in store—if not for their readers, then for their successors. "If Toussenel is to be believed," declaims Aronnax, the "plague" of yellow fever "is nothing beside that which will strike our descendants when the seas are emptied of whales and seals." The expectation in question—that the oceans are en route to becoming "huge centres of infection" (Verne 2019, 308)—relies on the extraordinary idea that whale and seal stomachs have been created by God to despumate (*écumer*) the sea surface. Thanks to overhunting, all this natural cleaning is coming to an end. The waters of the future, writes Toussenel, will be heavy with octopuses, jellyfish, and squid, as though heaps of cadavers were carried along by the waves. (This last, exceptionally macabre scenario may evoke sensationalizing images of the Ganges, where unconsumed human corpses were alleged to generate miasmatic cholera.) And these conditions will result in putrefaction on a universal scale, the appalling results whereof one strains to contemplate, let alone specify. Fortunately, writes Toussenel, he and his audience will by then be dead and buried (1868, 155–6).

A world depleted of whales and seals is on this account a diseased world, one that Toussenel imagines will see the populations of empires cut down by the sea wind as thoroughly as fire clears away young grass. To blame, ultimately, will be the policies of European and American governments which incentivize waging "war" on whales by offering unreasonable bonuses for their capture. The effect has been to encourage the extermination of creatures which have never done humanity the least wrong, as well as to deprive modern science any realistic chance of adequately describing them (Toussenel 1868, 156–9). The first of these complaints may recall Nemo's discomfiting calculus from *20TL*'s southern episodes, which figured baleen whales as objects of sympathy while deeming "cruel and malevolent" sperm whales "right to exterminate" (Verne 2019, 264). And the second adds yet another layer to the crises extinction is understood to occasion: Toussenel worries, here, over the awful eventuality that a species might disappear so rapidly as to preclude the possibility of its being registered in the annals of knowledge. Conseil's counsel against killing the Red Sea dugong springs to mind, as must its ineffectuality—and for all Aronnax's talk of important roles and ghastly fevers, the *Nautilus* will motor away from Suriname with "several tons" of manatee "meat" aboard (Verne 2019, 308). In the proceeding section, this problem and others jostle concerningly—but perhaps productively—with *20TL*'s persistent appeal for its conservation-minded inheritors.

Modernity, Disgust, and Xenophobia at Sea

"No words can describe the personal liberation that heading seaward bestows upon me. In this aquatic realm no man or woman is subject to the petty decrees of social bureaucracy." The quotation comes from the Canadian marine activist Paul Watson, cofounder of Greenpeace and Sea Shepherd, but the sentiment it expresses is quintessential Nemo (Khatchadourian 2007). Verne's antihero is a nationless rebel possessed of immeasurable wealth, matchless technology, and the self-image of a vengeful revolutionary. By the end of *20TL*, even Aronnax is unable to look past the righteous terror wrought by the *Nautilus* or its captain, whom he comes to regard as "no longer a fellow human, but a marine being, a spirit of the seas" (Verne 2019, 345). No student of Western environmental thought is unfamiliar with the clichés that human civilization is incompatible with nature's flourishing and that the path to better living is one best trod antisocially. Nemo and his intellectual fellow-travellers might be eligible, therefore, for dismissal as just a few more iterations of a well-known ideological type, were it not for the specifically and significantly oceanic aspects of their worldview—not to mention their continuing careers as cameo performers in the discourses of marine conservation science.

Just after its fleeting citation of Aronnax's whale-less "hotbeds of infection," Jackson's "Brave New World" ventures a "pollution thought experiment." "Try to imagine how disgusting the land would be," proposes the oceanographer, "if most of the garbage, sewage, and toxic waste of humanity had not been dumped into the ocean for the past several thousand years, and instead had been piled up on the land." With this image, Jackson is warming to an even more revolting, and important—"what I really want to talk about"—subject, which he calls "the rise of slime." Agricultural malfeasance and climate change are conspiring, he explains, to facilitate the "explosive growth of phytoplankton," the proliferation of "toxic dinoflagellate bloom[s]," and the resulting appearance of anoxic "dead zones" dominated by "jellyfish and bacteria" (Jackson 2010). The "rise of slime" and related concepts—like the fisheries scientist Daniel Pauly's "Myxocene" (2010, 61n)—influentially figure the existence and enlargement, over vast areas, of a particular kind of marine abject, scenes of extinctive absence that are defined simultaneously by repugnant presence (Quigley 2022). What is alarming about the brave new ocean, in other words, is not only that the ocean is becoming devoid of "all the species we like and the ecosystems we like" but that it is becoming literally suffocatingly full of "the stuff we don't want" (Jackson 2010).

"Climate change is going to be gross." The phrase—which heads a 2021 report in *The Atlantic* magazine on an expansion of mucilage ("snot") on the Sea of Marmara—neatly captures the distinctive conjuncture of speculative marine futures with feelings of disgust (Scatena 2021). Ursula Heise has observed that the predominant genres of "environmentalist storytelling" in general, and of extinction narratives in particular, have been tragedy and elegy (2010, 60). Offshore, those registers are joined—and sometimes supplanted—by something like (gross-out) horror. If this is related to a well-known teratological mode in imagining oceanic spaces and bodies, it is also unique. What is beheld, in visions like Jackson's, is the body *of* the sea becoming subject to an array of invidious pressures and metamorphosing, through the disappearance of benign species and ecosystems, into a monstrous engine of filth. Needless to say, the idea that the ocean is an entity susceptible to violations and (horrible) transformations is not new. What merits asking, nonetheless, is how a specifically modern sense for global—and above all marinal—economy interacts with, and generates, the spectre of waters growing unspeakably vile.

In *20TL*, that modern sense appears alongside growing concern for the character and consequences of modernity. Critics attentive to the environmental aspects of Verne's eighteen-volume *Voyages extraordinaires* (*Extraordinary Journeys*) figure them as persistently preoccupied by the unforeseen—and perhaps unforeseeable—impacts of technological change, industrial development, and the growth of consumer society. By these accounts, overhunting, (near-) extinction, and what we would nowadays call habitat loss are the leading emblems of a sustained vision of a society losing control of the instruments of civilization at the same time that it is triumphantly inventing them (Dupuy 2017, 9–16; Evans 1988, 93–4). If *20TL* is this worldview's pre-eminent expression, it also issues forcefully from *Le pays des fourrures* (*The fur country; seventy degrees latitude north*, 1873), where regions devoid of otters and beavers testify to the "massacres" carried out there by trappers who lack the sense even to spare young or pregnant animals (Verne 1873, 13–14). It comes forth, too, in by now familiar form, from *Le sphinx des glaces* (*The Sphinx of the Ice Fields*, 1897), where whales are reported to have become rarer and rarer in once-lively southern Seas (Dupuy 2014, 17).

If an act of speculation necessarily entails what DeCanio calls "the articulation of a moral vision," then in the *Voyages extraordinaires* that vision is informed substantially by the perceived downsides of industrial capitalist modernity (1994, 80). For environmental literary studies, this is nothing if not customary territory. The operations of *20TL*'s future-gazing may acquire a decidedly unfamiliar and uncomfortable look, however, once considered with a closer regard for certain parts of the novel's intellectual milieu. As we have seen, Aronnax's gory prefiguration of whale- and sealless hotbeds of infection has figured prominently amongst accounts of the novel's environmentalist prescience. For its source, Toussenel, the prospect partakes of a broadly anti-capitalist and anti-financialist spirit. Nineteenth-century "civilization" and its defining "commercialism" were, for the author of *L'Esprit des bêtes*, culpably involved in a host of modern aberrations, not least a decline in populations of France's native fauna and their replacement by comparatively ignoble species. Furthermore, environmental depredations such as these were for Toussenel the predictable outcomes of the growing influence of foreign mores, above all those of the English nation and—most troublingly, for present purposes—Jewish persons (Crossley 2004, 463–7).

Not long before the appearance of *L'Esprit des bêtes*, Toussenel had published *Les Juifs rois de l'époque: histoire de la féodalité financière* (*The Jews, Kings of the Era: A History of Financial Feudalism*, 1845). A socialist critique of mid-nineteenth-century commercialism in France and elsewhere, the text's analysis is also thoroughly anti-Semitic. Following the utopian social reformer Charles Fourier, who argued that modern civilization had dealt dreadful injury to the human passions, Toussenel comprehended nature as a mesh of correspondences interpretable in terms of their moral as well as scientific meanings. As Ceri Crossley has written, this entanglement of "zoology" with "social criticism" lent itself, for the author of *L'Esprit des bêtes*, to a hermeneutics—and a mode of speculating about the future—constituted to a substantial extent by "xenophobia." Recalling Toussenel's oeuvre here might seem a gratuitous foray into a peculiar and objectionable worldview, were it not for the fact that *Les Juifs rois de l'époque* proved a key text for the notorious anti-Semite Édouard Drumont and his late nineteenth- and early twentieth-century sympathizers (Crossley 2004, 460–3). To return to those images of heedless metropolitans, in *L'Esprit des bêtes*, while bearing all this in mind is to consider whom—precisely *whom*—their author may have had in mind.

Nemo, whose name is Latin for "nobody," exits *20TL* without betraying any substantive detail of his personal history or national identity. When he reappears in a subsequent novel, *L'île mystérieuse* (*The Mysterious Island*, 1874), he finally divulges the facts of his biography and the "immense disgust" for humanity—"the hatred for, and horror of, the civilized world"—that motivated his descent beneath the waves (Verne 1874, 569). In *20TL*, what is repellent about rapacious modernity takes shape through images of cephalopod and coelenterate bodies massing at the untended surfaces of future seas. Submerged amidst the heterogeneous host of influences that inform this spectacle are strains in mid-nineteenth-century quasi-conservationist thought that sully Aronnax's foresight with the taint of hate. Whether and in what manner Verne's present-day interlocutors ought to contend with stains of this sort is a question that will ultimately exceed the scope of this paper but that occupies its final paragraphs.

* * *

20TL's special status as an icon of oceanic consciousness is perhaps never more obvious than when it is deployed in a spirit of lapidary self-evidence. One very recent example comes from the journal *Marine Policy*, where several sentences from one of Nemo's paeans to his adopted home—"where man is never lonely, for he feels life stirring on all sides"—appear as the epigraph to an article on high seas governance (Pereira et al. 2023, 2). That status is, of course, really paradoxical. On the one hand, when bits of Verne's novel appear in places like *Marine Policy*, their currency as circulable, internationally acknowledged evocations of oceanity is briefly but potently confirmed. On the other hand, such treatments threaten to reconstruct *20TL* as primarily an artefact—better, jumble of artefacts—of citation, not a work endowed with the full array of complex compositional texture. This becomes a glaring problem when—as in the case of Aronnax's ventriloquism of Toussenel's foreshadowing—a favoured excerpt is found to be at least potentially freighted with ideological baggage that few twenty-first-century conservationists would care to carry, and at worst consequentially injurious to the integrity of its citers' projective claims. Would the discourses of the marine natural and social sciences (among others) be better off leaving literature well aside?

If *20TL* appears to offer some compelling justifications for answering this question in the affirmative, it may also furnish some important reasons for replying "no." For the critic Jacques Noiray, Verne's undersea adventure typifies a mode of integrating scientific knowledge and literary art that was giving way, by the 1870s, to a sense of those realms as categorically distinct (2007, 49). It is not necessary to accept Noiray's chronology of this divergence wholesale to recognize what is most suggestive about his analysis, which is that the literariness of recent attempts to figure ocean futures in general, and marine extinctions in particular, has histories. Those histories make available cautionary information respecting the genealogies of certain narratives, affects, and imageries. Beyond repositories of discomfiting complexity, however, they are spurs to hard and fruitful reading, thinking, and writing across practices and criticisms, towards not only "new stories" but more searching and more open accounts of how ecological concerns have taken shape and are shaping up (Solnit 2023). Neither a capitulation to imagination's instrumentalization nor a retreat into performative belletrism, such work would acquaint itself seriously with novels, whales, intellectual histories, extractive industries, and the oceanic worlds they have been making and might make. The sea demands as much.

References

Bellos, David. 1999. *Georges Perec*. London: The Harvill Press.

Bradbury, Ray. 1962. The Ardent Blasphemers. In *Jules Verne, 20,000 Leagues Under the Sea*, trans. Anthony Bonner, 1–15. New York: Bantam Books.

Bruijn, Iris. 2009. *Ship's Surgeons of the Dutch East India Company: Commerce and the Progress of Medicine in the Eighteenth Century*. Leiden: Leiden University Press.

Crossley, Ceri. 2004. Anglophobia and Anti-Semitism: The Case of Alphonse Toussenel (1803–1885). *Modern & Contemporary France* 12 (4): 459–472.

DeCanio, Stephen J. 1994. The Future Through Yesterday: Long-Term Forecasting in the Novels of H. G. Wells and Jules Verne. *The Centennial Review* 38 (1): 75–93.

Drabble, Margaret. 2014. Submarine Dreams. *New Statesman*, May 2–8, 2014, 48–51.

Dupuy, Lionel. 2014. L'écologie humaine: Petite histoire d'une revolution paradigmatique. *E-Working Papers em Ecologia Humana* 5: 1–21.

———. 2017. Jules Verne, l'homme et la terre: Une lecture écocritique des *Voyages extraordinaires*. *L'Esprit Créateur* 57 (1): 9–19.

Evans, Arthur B. 1988. *Jules Verne Rediscovered: Didacticism and the Scientific Novel*. Westport, CT: Greenwood Press, Inc.

Evans, Arthur B., and Ron Miller. 1997. Jules Verne, Misunderstood Visionary. *Scientific American* 276 (4): 92–97.

Heise, Ursula K. 2010. Lost Dogs, Last Birds, and Listed Species: Cultures of Extinction. *Configurations* 18 (1–2): 49–72. https://doi.org/10.1353/con.2010.0007

Jackson, Jeremy. 2008. Ecological Extinction and Evolution in the Brave New Ocean. *PNAS* 105 (Suppl. 1): 11458–11465. https://doi.org/10.1073/pnas.0802812105.

———. 2010. Brave New Ocean. *Smithsonian Ocean.*. https://ocean.si.edu/conservation/pollution/changing-tides-brave-new-ocean. Accessed 19 September 2023.

Khatchadourian, Raffi. 2007. Neptune's Navy: Paul Watson's Wild Crusade to Save the Oceans. *The New Yorker*, October 29, 2007. https://www.newyorker.com/magazine/2007/11/05/neptunes-navy. Accessed 2 April 2024.

Lawrence, Christopher. 1996. Disciplining Disease: Scurvy, The Navy, and Imperial Expansion, 1750–1825. In *Visions of Empire: Voyages, Botany, and Representations of Nature*, ed. David Philip Miller and Peter Hanns Reill, 80–106. New York: Cambridge University Press.

Maury, M.F. 1844. Remarks on the Gulf (Stream) and Currents of the Sea. *The American Journal of Science and Arts* 47: 161–181.

Noiray, Jacques. 2007. L'Inscription de la science dans le texte littéraire: L'exemple de *Vingt mille lieues sous les mers*. In *Jules Verne ou les inventions romanesques*, ed. Christophe Reffait and Alain Schaffner, 29–50. Amiens: Encrage Université.

Pauly, Daniel. 2010. *5 Easy Pieces: The Impact of Fisheries on Marine Ecosystems*. Washington, DC: Island Press.

Pereira, Laura M., Guillermo Ortuño Crespo, Diva J. Amon, Renuka Badhe, Salomão Bandeira, Frida Bengtsson, Miranda Boettcher, et al. 2023. The Living Infinite: Envisioning Futures for Transformed Human-Nature Relationships on the High Seas. *Marine Policy* 153: 1–17. https://doi.org/10.1016/j.marpol.2023.105644.

Quigley, Killian. 2022. Reading the Anthropocene Ocean. In *Anthropogenic Markers: Stratigraphy and Context*, ed. Christoph Rosol and Giulia Rispoli. Berlin: Max Planck Institute for the History of Science. https://doi.org/10.58049/ppqy-vd84.

Sala, Enric. 2015. Captain Nemo's Dream. *Architectural Digest* 85 (4): 106–109.

Scatena, Jenna. 2021. Climate Change is Going to Be Gross. *The Atlantic*, December 18, 2021. https://www.theatlantic.com/science/archive/2021/12/turkey-sea-snot-climate-change/620756/. Accessed 2 April 2024.

Solnit, Rebecca. 2023. "If You Win the Popular Imagination, You Change the Game": Why We Need New Stories On Climate. *The Guardian*, January 12, 2023. https://www.theguardian.com/news/2023/jan/12/rebecca-solnit-climate-crisis-popular-imagination-why-we-need-new-stories. Accessed 2 April 2024.

Toussenel, Alphonse. 1868. *L'Esprit des bêtes*. Paris: J. Hetzel.

Verne, Jules. 1873. *Le pays des fourrures*. Paris: J. Hetzel.

———. 1874. *L'île mystérieuse*. Paris: J. Hetzel.

———. 1978. *The Annotated Jules Verne: From the Earth to the Moon*. Translated and edited by Walter James Miller. New York: Thomas Y. Crowell.

———. 2019. *Twenty Thousand Leagues under the Seas*. Translated by William Butcher. Oxford: Oxford University Press.

Worster, Donald E. 1977. *Nature's Economy: The Roots of Ecology*. San Francisco: Sierra Club Books.

CHAPTER 4

Narwhals for All Seasons: Representation, Evasion and Absence

Rick De Vos

Narwhals in Winter

Autumn in the Arctic Ocean brings a rapid shortening of days, as the Sun begins to recede and temperatures decrease. The Arctic sea-ice begins to refreeze and become covered in snow. These changes provide signals for narwhals from the Baffin Bay population to move east and south, dispersing into Disko and Baffin Bays, and the Davis Strait (Heide-Jørgensen 2010). Familial pods of narwhals make their way along the western coast of Greenland, arriving between late October and early November. Each group travels to their traditional wintering area. These areas have distinct and specific characteristics: they are deep, off-shore and covered in ice. Swimming under the moving pack-ice, the narwhals dive to depths of up to 2000 metres in order to catch Greenland halibut, Arctic cod and armhook squid, returning to the surface to breathe through cracks and leads in the ice (Laidre and Heide-Jørgensen 2005a, 510).

R. De Vos (✉)
Curtin University, Perth, WA, Australia
e-mail: R.DeVos@curtin.edu.au

© The Author(s), under exclusive license to Springer Nature Switzerland AG 2024

V. Fibisan, R. Murray (eds.), *Blue Extinction in Literature, Art, and Culture*, Palgrave Studies in Animals and Literature,
https://doi.org/10.1007/978-3-031-69910-8_4

For the five months that they winter here each year, the narwhals live in almost total darkness. Their wintering habitat is of the highest importance to them for their physical and social wellbeing: these are the months in which narwhals feed most intensely, building up fat to sustain them during their spring migration to the North Water and the bays and fjords of north-western Greenland, when they will travel more than 1000 kilometres over a two-month period. They rely on body reserves accumulated in winter to sustain their year-round activities including, for female narwhals, calving and nursing in the late spring and early summer (Chambault et al. 2023). In years where the ice cover is most extensive, the narwhals will swim further and faster (Laidre and Heide-Jorgensen 2005b, 52–3). Their bodies have evolved to withstand the extreme cold, the extreme water pressure and the darkness. Echolocation via high frequency clicks is utilised for spatial orientation and locating prey, and complex combinations of higher frequency clicks, whistles and pulsed tones are used for communicating with each other. These vocalisations have been recorded as increasing in relation to sea-ice concentration (Marques 2021). Darkness, deep water, extensive pack-ice cover with small holes and cracks for breathing, relative quiet and a plentiful supply of halibut, squid and cod, combined with a close connection to their familial pods and larger groups, all contribute to narwhals' ideal feeding and wintering conditions. As winter is a long season in which narwhals are not visible to humans, it is a time that has led to diverse speculation and imagination regarding their activities, relationships and identities. Despite the fact that they spend nearly half of the year in this submerged/off-shore state, however, the importance of winter living and habitats for narwhals is often forgotten in their representations. Indeed, cultural and scientific representations of narwhals, limited mainly to summer observations and encounters, run the risk of misrepresenting the agency, complexity and diversity of narwhal lives, in turn contributing to the endangerment of those lives, and ignoring and mitigating ways of conserving them.

Mammals of Least Concern

The North Atlantic Marine Mammal Commission (NAMMCO) and the International Whaling Commission (IWC) are two international bodies that monitor whale populations and provide data regarding the conservation of marine mammals, including narwhals. The member nations of NAMMCO, Greenland, Faroe Islands, Iceland and Norway, are all currently involved in whaling activities. Although the IWC is a larger body with a more explicit conservation mandate, both commissions use the word "stocks" in discussing cetacean numbers and populations, including those of narwhals, and their recommendations focus on the level of harvesting, from both commercial and Indigenous subsistence hunting. While professing a commitment to cetacean conservation, a more proprietary relationship with narwhals is expressed. Both bodies have chosen not to support bans on hunting narwhal, even in areas of East Greenland where population numbers have decreased significantly, calling instead for reduced quotas. The combined knowledge representing narwhal conservation status and narwhal futures has contributed to narwhal populations being declared "Of Least Concern" by the International Union for Conservation of Nature (IUCN) (Lowry et al. 2017). The IUCN has stated that narwhals do not meet their criteria for threatened categories due to their total population numbers, while conceding that some subpopulations warrant further assessment.

Narwhals have endured and thrived in their sea-ice environment for at least the past one million years, living in familial pods and following their established seasonal migration patterns within the Arctic Ocean. These patterns of movement are tied to the dynamics of sea-ice, where and when it forms, and where and when it melts. They return to their same winter habitat every year regardless of the state of the sea-ice. The shifting dynamics of the Arctic sea-ice regime have, for the past three decades, however, produced an environment that has become less predictable. Through most of the Arctic Ocean, the sea-ice has decreased both in area and in thickness as a consequence of atmospheric and oceanic warming. In turn, there has been an increase in rates of carbon dioxide fixation in ice-free areas (Notz and Marotzke 2012). The Baffin Bay and Greenland Sea sectors of the Arctic indicated a trend of increasing sea-ice between the middle and latter part of the twentieth century, a phenomenon attributed to these sectors' relatively deep-water areas, which are believed to limit nutrient re-supply in spring and summer (Brown and Arrigo 2012). This rate of sea-ice increase has started to plateau and reverse, however, since the beginning of the twenty-first century, with all Arctic Ocean sectors now showing severe ice loss in summer. Reduced sea-ice cover in the Arctic has the potential to drastically alter a critical seasonal habitat supporting high concentrations of marine mammals. In addition to its important role as a platform for marine mammals and birds, sea-ice serves as a habitat for a unique, highly specialised community of bacteria, algae, protozoans and metazoans, which contribute to the biogeochemical cycles of polar seas (Laidre et al. 2008, S98–9; Horner et al. 1992). Sea-ice cools the water column in spring, when herbivorous zooplankton and copepods graze on phytoplankton. Reduced sea-ice cover can result in a premature phytoplankton bloom, meaning that the phytoplankton and algae may die before copepods and amphipods ascend to graze on them. This rupture in trophic coupling may then lead to a cascading breakdown in feeding hierarchies, as Arctic cod and larger predators face a limited food supply.

Exacerbating the problem of fish stock depletion for narwhals is the increasing efficiency of fishing operations in the area and the emerging deep-water fishery in the Baffin Bay and Davis Strait waters, which has started taking increasing amounts of Greenland halibut each year, as well as other deep-sea fish species favoured by narwhal (Laidre et al. 2004, 430). Recent research has indicated decreasing abundance of halibut and bycatch species in these waters, as well as decreases in the size and weight of fish caught (Jørgensen et al. 2014, 850). A 2008 study concluded that due to their narrow distribution, specialised feeding habits and seasonal dependence on sea-ice, narwhals appeared to be the Arctic marine mammal species most sensitive to climate-induced habitat change (Laidre et al. 2008, S112–4). The widespread melting of glaciers and sea-ice, and the trends of dramatically increasing Arctic temperatures, shortening winters and reduction in snow, all directly affect the habitat and lives of narwhals. While they appear to live in remote areas far from metropolitan centres, the habitat changes they are experiencing are for the most part anthropogenic. Hunting, pollution, seismic exploration and increased shipping and tourism, added to the consequences of summer sea-ice loss, increased carbon dioxide fixation and raised atmospheric and oceanic temperatures in the Arctic, all raise questions about the narwhals' continued survival, and whether the Arctic sea-ice will continue to sustain them.

A report published in 2021 expressed grave concerns regarding the future of a narwhal population in Scoresby Sound, Southeast Greenland (Garde et al. 2022). The report asserted that excessive and opportunistic hunting had led to a decreased proportion of females, a decreasing trend in pregnancy rates, an overrepresentation of old males and a lack of calves and juveniles in the group. The trends pointed to the targeting of the slower members of the group. These results suggested an irreversible population decline. Despite this indication that specific populations and subpopulations can run the risk of endangerment, narwhals continue to be represented and made sense of as a species of least concern.

Cultural Representations

What distinguishes narwhals most visibly from all other cetaceans is the spiralled tusk found most commonly on adult male narwhals. Adult narwhals have two elongated teeth, and in adult males the left tooth continues to grow and protrude through their jaws, developing into tusks that can reach up to 3 metres in length. Female narwhals sometimes have a small tusk, males occasionally have no tusk and narwhals with two tusks are not uncommon (Heide-Jørgensen 2018). These tusks are salient features in Inuit and Western cultural and scientific representations of narwhals, and a recurrent concern of these representations is to make sense of them.

Narwhals have been the subjects of Inuit stories and art for hundreds of years. Inuit people living in Greenland and in Nunavut and the Northwest Territories in Canada have distinct histories and ways of understanding and representing narwhals, but common elements can be identified despite this diversity of languages and cultural experiences. The context in which narwhals are viewed is that of animals who are connected to Inuit communities, and highly prized by them. They are seen as an important source of food, as well as raw materials, but are recognised as elusive, shy and difficult to catch. Narwhal tusks have remained a focus of both visual art and folktales.

While it has a number of variants, there is a shared Inuit story of how the narwhal came to exist. A common element in this creation story is that a woman treated her blind son deceitfully and cruelly over a sustained period of time. One day they went fishing and the son persuaded his mother to tie the end of a harpoon rope around her waist. When he threw the weapon and successfully struck a large beluga whale—which he had deliberately aimed to hit—she was pulled into the ocean. There, in the dark depths, her long hair began to twist around the rope, hardening and forming a twisted tusk, and thus she became a narwhal (McLeish 2013, ix–x; Boissonneault 2012, 92). This movement of life from land to sea, and from the surface to the depths, is represented as both physical punishment and spatial banishment. The narwhal is cast as female, cruel and deceptive, deserving of her fate. It is the tusk and its formation that changes the identity of the woman from human to narwhal, with the narwhal connected to both humans and beluga. The tale also positions the narwhal as being created after humans and beluga.

The contrast between this creation story and the experiences of Inuit people who hunt and consume narwhals is stark and immediate. The story of the mythical first narwhal is told and listened to in Inuit communities with a clear picture of what narwhals look like, what they taste like and what it means to be in the presence of narwhals. The relationship is both a cultural and an instrumental one. Hunting narwhal involves a set of very specific knowledges, experiences and practices that are passed down through generations. The consequences of climate change and the emergence of other professional choices for younger Inuit community members, however, mean that these skills and experiences are possessed by fewer people.

Narwhals and narwhal stories have been the subject of traditional and contemporary Inuit art, including soapstone, serpentinite, whalebone and wooden carvings and sculptures, and in paintings, engravings, stencils and prints. Paintings and sculptures by local Inuit artists are sold directly to tourists via cruise ship stalls in West and East Greenland, while a larger number are sold to art galleries in Canada, the United States and throughout Europe. Only rarely are Inuit artists given individual recognition in international art markets, and rarely do they have economic control of their work (Nixon 2020).

Representations of narwhals in Europe have been entangled with those of other marine creatures, both mythical and actual, with stories and images generated from a paucity of real and imagined encounters between narwhals and European sailors. These encounters led to descriptions of creatures that were sometimes monstrous, sometimes alluring, sometimes ghostly or ethereal. Amongst the mythic sea creatures dominating the Swedish writer and cartographer Olaus Magnus's influential illustrated 1539 woodcut *Carta Marina* is an image of a fierce "sea unicorn," with a large horn protruding from the sea monster's head (Nigg 2013). Conrad Gesner's *Icones Animalium*, first published in 1553, also includes a monstrous "sea monoceros," its horned head emerging from the waves (Nigg 2013). Depictions and descriptions like this endured to the extent that the existence of narwhals as real animals remained shrouded in doubt, and even up to the twentieth century many people were unaware of narwhals. In particular, European representations of narwhals have been conflated with and overshadowed by tales and images of unicorns, legendary equine, bovine or caprine land-dwelling animals with a single spiralled horn emerging from their foreheads. Artistic and literary representations of unicorns have flourished for over a thousand years. Inuit and Norse settlers in

Greenland traded with each other since at least the ninth century AD (Dectot 2018; Pluskowski 2004). These exchanges led to Norse settlers developing a trade in narwhal tusks, selling them to other European traders as unicorn horns. The desire for unicorns, for their horns and for their actual existence, helped to place great monetary value on these tusks. The multitude of tusks to be found in European royal palaces, the homes of the wealthy and later, in State museums attests to the association of unicorns and their horns with social status and regality. Despite the efforts of traders to mystify the origin of their valuable commodities, by the seventeenth century writers had identified narwhals as the source of the purported unicorn horns. This did not diminish the fascinations for unicorns, however, with narwhals defined and represented in relation to unicorns, as "sea-unicorns."

Public representations and discussion of narwhals in the West have rarely omitted the enduring association with unicorns, often carrying an assumption that this connection adds to the appeal of narwhals. Narwhal tusks are displayed in the Unicorn Tapestry Room at the Met Cloisters in New York, in the Victoria and Albert Museum in London and in the Cluny Museum in Paris, in each case as part of unicorn-related exhibitions. Marie-France Boissonneault has argued that the narwhal-unicorn connection has contributed to the emblematic status of narwhals in Canada (2012, 92–94). Unicorns were depicted in Celtic mythology, and a unicorn representing Scotland is displayed on the British coat of arms. Narwhals appear on the coat of arms for the Northwest Territories and Nunavut, provinces encompassing the Canadian Arctic. Narwhals have appeared on Canadian and Greenlandic postage stamps. The narwhal's name and image have been utilised by hotels, bars, breweries, clothing boutiques and manufacturers, arts and craft stores, computer software and a Canadian independent news website with an environmental focus. In contrast to the Inuit creation story, these images invariably depict narwhals as singular and male.

Narwhals' cultural association with unicorns and their emblematic status suggests a shifting of narwhal images away from marine contexts to terrestrial ones. Unlike many Inuit representations, Western images are more likely to be produced in the absence of any experience or knowledge of narwhals. The focus on tusks also implies a focus on male narwhals as more significant. The anthropomorphisation of narwhals in popular culture can be seen as adopting and extending these shifts in commercial contexts.

Narwhals have proved to be a popular subject of children's picture books and story books. Within such titles there is a consistent practice of re-articulating narwhal bodies by shifting their tusks to the centre of their foreheads, suggesting a marine unicorn. The moving of the tusk allows the anthropomorphised narwhal to display a smiling face and allows for other expressions. The whales are thus rendered benign, and strange but knowable. The smiling face is an image that is utilised extensively in pictures illustrating these stories. Photographs of beluga whales depicting the upturned edges of their mouths are often interpreted as displaying smiles, and these images might motivate the portrayal of narwhals as smiling, as well as addressing the problem of inscrutability posed for humans by narwhals' elusive faces. Narwhals in popular culture are often positioned as lovable, innocent, unsophisticated and vulnerable. At times, however, they are portrayed as possessing some latent power or ability, the source of which is invariably the narwhal's tusk. Ben Clanton's *Narwhal and Jelly* book series is designed for young children (2019). While set in an imagined undersea world the main characters, a narwhal and a jellyfish, engage in human-style adventures. *Wendell the Narwhal* by Emily Dove (2016) presents the story of a young narwhal who wants to make music and ends up being a conductor, employing his tusk as a baton. Lisa Mantchev's *Someday, Narwhal* (2017) follows the fortunes of a pet narwhal who lives in a fishbowl and longs for adventure. Joseph Pro's *The Adventures of Miki the Narwhal* (2012) presents a young narwhal's quest for adventure and how his father's advice helps him escape a dangerous situation. *Not Quite Narwhal* by Jesse Sima (2017) is the story of a young unicorn who was born under the sea to a family of narwhals.

All of these publications present the tusk as a horn in the middle of the character's forehead. None of them specifically engage with the Arctic Ocean or with sea-ice. They depict singular creatures, negotiating their difference from other creatures, rather than social life with other narwhals. The stories portray anthropomorphic narwhals who can see and hear like humans, are autonomous as individuals and form social bonds with other species. The singularity of the narwhal characters allows for the narratives to shape the characters' motivations and their fates in ways that would not be possible in narwhals' more social contexts and lives. In this sense the idea of narwhal as specimen takes precedence over that of species. A more realistic adventure is depicted in *Narwhal: The Arctic Unicorn* by Justin Anderson (2023), where a pod of narwhals migrates north as winter ends and spring begins. The story describes the pod's search for food, their avoidance of a predatory polar bear and how they navigate the sea-ice. The story is supplemented by background notes about narwhals. The tusks of the male members of the pod are depicted in their correct physical place, and the collective, social responses of the narwhals are described.

Plush toys, fluffy clothing approximating narwhal costumes, T-shirts and other merchandise represent narwhals as cute, cuddly, affectionate and endearing. Greeting cards, posters and other objects displaying narwhal motifs affirm and celebrate narwhal difference as a quality that might be shared by the owner or wearer. This transfer of meaning might be seen as operating in a similar way to representations of unicorns. Irony is utilised in the Avenging Narwhal play set ("with four magical tusks and three adorable animals to impale"), where a rubber narwhal's (detachable) tusks are re-imagined as weapons of vengeance to be deployed against a baby seal, a penguin and a koala.

These popular culture representations consistently convey a desire to identify with or connect with narwhals, to separate them from their habitat and biotic relationships, to move them from sea to land and to reposition their tusks as horns. Cultural desire for narwhals is closely associated with desire for unicorns, where narwhal representations take on a mythological quality. In doing so they move away from biological and environmental discourses, and into the realm of fantasy and magic.

Scientific Representations

Western cultural and scientific representations inform and reinforce each other. The appearance and anatomy of narwhals, the significance and value of the tusks of male narwhals and speculation regarding their undersea relationships and movements are common concerns of both Western science and culture. Western scientific research since the early twentieth century has principally been concerned with the morphology and taxonomy of narwhals, in particular the function of tusks and those features that distinguish narwhals from other cetaceans. Diagrams derived from hand-drawn sketches have been used to represent these observations and speculations. Other concerns have included the trade in narwhal tusks, narwhal hunting, as well as seasonal movements and behavioural observations. Most narwhal research has taken place, however, in the past 20 years, undertaken by a relatively small number of researchers, most of whom have been based at and supported by the Greenland Institute of Natural Resources. The Institute, a body advising the Greenlandic government and aligned with NAMMCO, has the stated objective of providing the scientific basis for "an assessment of sustainable use of the living resources in and around Greenland as well as protecting the environment and securing the biological diversity" (Pinngortitaleriffik n.d.).

Recent research has helped to establish that narwhals can be seen as belonging to three distinct populations with distinct migratory patterns and discrete summering and wintering locations. The research has addressed the movements and behaviours of specific narwhal populations with the use of enhanced surveillance technologies, including aerial surveys, infra-red imagery, wavelength photography and satellite-linked time-depth-temperature recorders. Articles and reports communicating this research make use of maps and graphs to illustrate these findings. Only a limited number of photographs depict a pod of narwhals surfacing, and these tend to be reproduced many times in scientific publications as well as in magazines and news articles. Each of these photographs and survey images depict narwhal movements in summer. While an extensive and diverse range of data has been produced, narwhal familial relationships, narwhal communication and the reasons for their open-water sea-ice affinity continue to evade clear explanation.

One prominent researcher, Dr Kristin Laidre, commented on problems associated with narwhals' cryptic nature:

> The narwhal is probably one of the worst study animals you can choose. They spend most of their life underwater at very deep depths. You have to travel several hundred kilometres offshore in dense sea ice to get to them. The whales are difficult—they're really shy and easily scared away. They can sneak right past you and you won't even see them. (Kennedy 2012)

This brief quote crystallises narwhal scientific representation, in particular the need to conceive of narwhals as knowable, and to present research as conclusive. Their elusiveness is seen as a problem rather than a strength, a problem for those choosing to undertake narwhal research. Noise is seen as related to narwhal sensitivity rather than narwhal perception and surveillance. Pack-ice might indeed be a way of increasing and facilitating narwhal echolocation abilities and parameters by providing hard structures as well as surfaces that conceal narwhal movements from human and other predators.

In recent years Inuit traditional ecological knowledge, based on Inuit people's longer and more immediate experiences and observations of narwhal, has both challenged and complemented Western scientific research on narwhals. Aspects of this knowledge and research have been published from case studies conducted by Canadian social scientists (Pearce 2015). The felt realities of climate change in the Arctic have motivated Inuit communities to increase their participation in, and control of, environmental research. Narwhals have been a subject of specific interest to communities because of the cultural and economic significance of hunting narwhals, their importance as a food source and the value of narwhal tusks as a supplementary source of income.

An interview with the late Kooneeloosee (Cornelius) Nutarak OC, a Mittimatalik (Pond Inlet) elder, artist and Inuit cultural knowledge holder, in 2006, provided a salient example of how Inuit knowledge of narwhal lives was shaped by generations of encounters, observations and close study (ELOKA 2010; Nweeia et al. 2009). The interview, one of a number conducted by the American marine mammal scientist and dental anthropologist Martin Nweeia, focused on narwhal tusks, but also provided new insights regarding narwhal pods and familial groups, and the classification of narwhals and possible subspecies. Nutarak described tusks as diverse in their shapes, colours and density, and differentiated between males in terms of colour and size, and how these indicated different roles within migrating pods. He also discussed their ability to sense orcas, and how narwhals quickly dived to greater depths to avoid these predators. Like other Inuit hunters, Nutarak expressed the belief that narwhals were more plentiful than they appeared, noting that previous declines in observed numbers had been followed by periods of recovery, and that hunting need not be reduced or curtailed. He had observed that motorboats and shotguns made hunting more difficult, in contrast to kayaks and harpoons, because of narwhals' ability to detect noise at great distances.

Savssats and Blessings

Narwhals are regularly killed in ice entrapments on their wintering grounds, when sudden temperature drops and changes in pressure conditions produce a rapid formation of fast ice in bays and passages used by the narwhals. The leads where narwhals surface to breathe freeze over quickly, leaving them stranded without enough breathing holes to guide them to open water. The narwhals squeeze into shrinking openings in the solid ice, with many suffocating and dying while fighting for air in these tight spaces. Others become easy prey for polar bears who gather around the breathing holes. Researchers have categorised these winter entrapments as a source of natural mortality for narwhals (Heide-Jørgensen et al. 2002).

Local Greenlanders have been aware of these entrapments for centuries, describing them as *savssats*, which translates approximately to "something that is served up." Accounts of these events represent them as fortunate ones. The collective noun for a group of narwhals in English is a "blessing," and given the rarity of seeing such a group, a connection with savssats may be conjectured. If savssats are detected, they provide an opportunity for hunters to shoot and harvest the trapped narwhals. Most historical entrapments have been reported in narwhal wintering areas, especially around Disko Bay, West Greenland, where the largest documented entrapment of over 1000 narwhals occurred in 1915. Over the past 15 years, however, a series of ice entrapments have been reported or observed in narwhal summering grounds, at a time when narwhals were not expected to have been in the locality (Laidre et al. 2012). At one such entrapment, in late November 2009, approximately 50–100 narwhals were trapped in Inglefield Bredning, approximately 15 km southwest of Qaanaaq. In February 2010, between 30 and 100 narwhals were found trapped in fast ice 5500 metres from Herbert Island in the Inglefield Bredning, 10 km from the November entrapment. In each case the entrapments were observed and reported by Inuit hunters and, after the situation was determined to be irreversibly lethal, management authorities gave permission for the narwhals to be harvested for subsistence purposes.

In each of the recent entrapments, it appeared that narwhals had delayed leaving their summering areas until winter, rather than leaving in early autumn. As autumn freeze-ups have shown a delaying trend each year, these entrapments might suggest a trend in prolonged summer residence time as narwhals adapt to a longer open water season. However, just as likely is the possibility that the narwhals had remained in their summering grounds in response to seismic exploration activity taking place in northern Baffin Bay between September and October 2009. Seismic airgun surveys, in search of offshore oil and gas deposits, are conducted by trawling arrays of airguns behind a ship which explode every 10–15 seconds. The noise permeates the seafloor and the signal that reflects back is deciphered to portray the substrates below. These surveys can often be heard thousands of miles from the source and are known to either kill marine mammals and fish, or cause hearing and echolocation degradation as well as habitat displacement, stress and migratory disruptions

(Heide-Jorgensen et al. 2013b, 53). Seismic exploration can now be seen as serving up these entrapments/*savssats*, or blessings.

Narwhals in Museums

Just as the tusks of male narwhals have been incorporated into museums as part of unicorn-related exhibits, the displaying of narwhals as preserved and taxidermized specimens reveals the way that the practices of cultural and scientific representations combine in reinforcing specific meanings. Preserved specimens are displayed in many museums around the world, often in contexts related to Arctic environments and the effects of climate change. The killing of narwhals, and the often-violent histories of trade and conquest that produced narwhal bodies and tusks, are sublimated in the display of singular narwhal bodies. The displayed body represents a discrete species evidenced by a single specimen, one differentiated from other cetaceans, other Arctic denizens and other species within the vicinity of the specimen. The discrete tusks that are perhaps displayed in another area of the museum, or in other museums or art galleries, also hint at this violent history but are instead recast as the objects of an artistic history removed from this violence.

Steve Alsop describes a visit to the Royal Ontario Museum in Toronto, Canada, and his encounter with a preserved narwhal specimen, within the context of science education and affective encounters (2018). He describes navigating large numbers of schoolchildren in making his way to the exhibit, a group of whom are singing "The Narwhal Song," a song from a popular children's television programme. The specimen proves to be large and impressive, but difficult to take in from a single perspective. Alsop reflects on the way inanimate stuffed individuals are utilised in presenting a sense of a materially embodied presence. Taking account of the way narwhal historical accounts are entangled with unicorn legends, Alsop works to separate the meanings narwhal generates in popular settings:

> I ask myself, are there differences between encountering a narwhal as a unicorn or a whale? I think there are. The affections of mystery and wonderment seem to fade (or at least become secondary) when the narwhal becomes *Monodon monoceros*, a medium-sized, pale-coloured whale with a large 'tusk'. The physical demise of this narwhal is simultaneously the birth of a universalised scientific object. (Alsop, 59)

Discourses of marine mammal biology and taxonomy, Arctic ecology, sustainable hunting, children's stories and entertainment, and commerce and merchandising coalesce in presenting museum exhibits that portray narwhals in ambiguous ways, as lovable, as impressive, but also removed, remote and mysterious.

Narwhals in Summer

By late March, solar radiation in the Arctic increases as the days lengthen, marking the onset of the spring phytoplankton bloom. The larvae and juveniles of benthic animals move to the coastal fast ice to feed on the ice algae, along with pelagic copepods and amphipods, tiny crustaceans favoured by Arctic cod, in turn the favoured prey of Greenland halibut. Between March and May, as the narrow leads in the ice open up into channels, narwhals undertake their spring migration from the open waters of Baffin Bay and Disko Bay toward and along the coast of West Greenland, feeding on Arctic cod and halibut along the edge of the sea-ice, before gathering in the North Water (Heide-Jorgensen et al. 2013a, 331–2). This is a time of mating and breeding for narwhals, with gestation lasting about 14 months. They make their way into the bays and fjords of northwestern Greenland and the Canadian Arctic Archipelago, with a large number entering Inglefield Bredning, where they spend the summer.

Representations of narwhals invariably assume and utilise the Arctic summer as the time of narwhals, the time depicted in photographic images of narwhals. Narwhals continue to spend a great deal of their time in summer further down the water column, displaying a clear preference for colder temperatures. However, narwhal activity near the surface combined with the lack of sea-ice provides visibility that carries with it prospects for human practices: for hunting, for studying and monitoring narwhals, and for viewing and photographic opportunities for Arctic tourists. These prospects, however, do not extend favourably to narwhals. Narwhal surface activity does not necessarily signify narwhal flourishing: it can also indicate extended periods of vulnerability and disorientation.

One aspect of the narwhal problem is that we think we know narwhals because we can imagine, depict and narrate them, and use them to generate cultural meanings. We also think we know them because we can collect or refer to data regarding narwhal observations. However, narwhals present us with a more daunting problem: that of evasion, avoidance and absence. They do not seek to connect or communicate with humans, nor do they express the desire to derive benefit from us. Narwhals do not survive in captivity, are not drawn to human activity or enticement and have not overtly modified their behaviour in response to anthropogenic changes to their environment, other than to double down on avoidance and evasion. Why they are so highly philopatric, returning to their same winter habitat every year regardless of the state of the sea-ice, remains a mystery. While the shelter of the densely packed ice might provide some protection from orcas and humans, and the deep waters under the ice might present a reliable supply of halibut and squid, narwhals' affinity for sea-ice exceeds these instrumental functions, evading our understandings of home, habitat and history. Seeking silence, darkness, ice, deep water, deep-water fish and squid, narwhals avoid and evade humans, whose activities and approaches denote danger and death.

Narwhals' choices suggest that the sea-ice constitutes a cultural space and a cultural time as much as a functional habitat. While the activities of mating, calving and detecting and avoiding predators, activities involving the negotiation of difference are prime concerns of narwhals in summer; narwhals in winter become hunters, both individually and together. It is a time of greater familial communication and connection and less vulnerability, a time for accumulating knowledge: how to find and hunt food, how to locate cracks and leads in the ice, how to negotiate space in the darkness, how to locate each other, how to speak and listen and how to forge connections that will allow pods to migrate together as autumn approaches. Just as importantly, the practices of watchfulness, avoidance and evasion, practices that have served narwhals for at least one million years, continue to be passed on even as new threats emerge. This extremely distinctive set of choices and practices, which includes their refusal to live in captivity, restricts human knowledge of narwhals to their summer activities and appearances.

While real narwhals continue to remain hidden for months each year under the Arctic pack-ice, albeit a cover that is measurably decreasing, cultural representations re-articulate narwhals as tusks, as sea unicorns, as skin and blubber, as hunting trophies, as harvestable commodities, as loveable mythological characters and sometimes even as creatures of vengeance. Such images conceal the threat of endangerment and extinction, and more importantly the narwhal times and spaces that constitute a threat to their survival. It can be argued, then, that knowing and loving narwhals is leading to their endangerment. By refusing their watery ontology and implicating them in our lives, we are creating a false intimacy with a group of animals clearly asking to be left alone, and for their time and space to be respected.

References

Alsop, Steve, and Justin Dillon. 2018. Encounters with a Narwhal: Revitalising Science Education's Capacity to Affect and Be Affected. In *Navigating the Changing Landscape of Formal and Informal Science Learning Opportunities*, ed. Deborah Corrigan et al. Cham: Palgrave. https://doi.org/10.1007/978-3-319-89761-5_4.

Anderson, Justin. 2023. *Narwhal: The Arctic Unicorn*. Sydney: Walker Books Australia.

Boissonneault, Marie-France. 2012. Beauty and the Enchanted Beast: The Narwhal (Monodon monoceros) in the Canadian Cultural Landscape. *Tamkang Review* 42 (2): 87–110.

Brown, Z.W., and K.R. Arrigo. 2012. Contrasting Trends in Sea Ice and Primary Production in the Bering Sea and Arctic Ocean. *ICES Journal of Marine Science* 69 (7): 1180–1193.

Chambault, Philippine, Susanna B. Blackwell, and Mads Peter Heide-Jørgensen. 2023. Extremely Low Seasonal Prey Capture Efficiency in a Deep-Diving Whale, the Narwhal. *Biology Letters* 19 (2): 20220423.

Clanton, Ben. 2019. *Narwhal: Unicorn of the Sea!* New York: HarperCollins.

Dectot, Xavier. 2018. When Ivory Came from the Seas. On Some Traits of the Trade of Raw and Carved Sea-Mammal Ivories in the Middle Ages. *Anthropozoologica* 53 (1): 159–174.

Dove, Emily. 2016. *Wendell the Narwhal*. Vancouver: Simply Read Books.

ELOKA. 2010. *Cornelius Nutarak Interview. Exchange for Local Observations and Knowledge of the Arctic*. University of Colorado, Boulder. https://eloka-arctic.org/communities/narwhal/interviews/nutarak.html. Accessed 2 April 2024.

Garde, Eva, Outi M. Tervo, Mikkel-Holger S. Sinding, Nynne H. Nielsen, Claus Cornett, and Mads Peter Heide-Jørgensen. 2022. Biological Parameters in a Declining Population of Narwhals (Monodon monoceros) in Scoresby Sound, Southeast Greenland. *Arctic Science* 8 (2): 329–348.

Heide-Jørgensen, Mads Peter. 2018. Narwhal: Monodon Monoceros. In *Encyclopedia of Marine Mammals*, ed. Bernd Würsig, J.G.M. Thewissen, and Kit M. Kovacs. Cambridge, MA: Academic Press.

Heide-Jørgensen, M.P., P. Richard, M. Ramsay, and S. Akeeagok. 2002. Three Recent Ice Entrapments of Arctic Cetaceans in West Greenland and the Eastern Canadian High Arctic. *NAMMCO Scientific Publications* 4: 143–148.

Heide-Jørgensen, M.P., K.L. Laidre, M.L. Burt, D.L. Borchers, T.A. Marques, R.G. Hansen, M. Rasmussen, and S. Fossette. 2010. Abundance of Narwhals (Monodon monoceros) on the Hunting Grounds in Greenland. *Journal of Mammalogy* 91: 1135–1151.

Heide-Jørgensen, M.P., P.R. Richard, R. Dietz, and Kristin Laidre. 2013a. A Metapopulation Model for Canadian and West Greenland Narwhals. *Animal Conservation* 16 (3): 331–343.

Heide-Jørgensen, Mads Peter, Rikke Guldborg Hansen, Kristin Westdal, Randall R. Reeves, and Anders Mosbech. 2013b. Narwhals and Seismic Exploration: Is Seismic Noise Increasing the Risk of Ice Entrapments? *Biological Conservation* 158: 50–54.

Horner, Rita, Stephen F. Ackley, Gerhard S. Dieckmann, Bjorn Gulliksen, Takao Hoshiai, Louis Legendre, Igor A. Melnikov, William S. Reeburgh, Michael Spindler, and Cornelius W. Sullivan. 1992. Ecology of Sea Ice Biota: 1. Habitat, Terminology, and Methodology. *Polar Biology* 12 (3): 417–427.

Jørgensen, Ole A., François Bastardie, and Ole R. Eigaard. 2014. Impact of Deep-Sea Fishery for Greenland Halibut (Reinhardtius hippoglossoides) on Non-Commercial Fish Species off West Greenland. *ICES Journal of Marine Science* 71 (4): 845–852.

Kennedy, Caitlyn. 2012. The Narwhal's Tale: Surviving Sea Ice Change. *NOAA Climate.gov*. https://www.climate.gov/news-features/features/narwhals-tale-surviving-sea-ice-change. Accessed 2 April 2024.

Laidre, Kristin L., and Mads Peter Heide-Jørgensen. 2005a. Arctic Sea Ice Trends and Narwhal Vulnerability. *Biological Conservation* 121: 509–517.

———. 2005b. Winter Feeding Intensity of Narwhals (Monodon monoceros). *Marine Mammal Science* 12 (1): 45–57.

Laidre, K.L., M.P. Heide-Jørgensen, O.A. Jorgenson, and M.A. Treble. 2004. Deep-Ocean Predation by a High Arctic Cetacean. *ICES Journal of Marine Science* 61 (3): 430–444.

Laidre, Kristin, Ian Stirling, Lloyd F. Lowry, Oystein Wiig, Mads Peter Heide-Jørgensen, and Steven H. Ferguson. 2008. Quantifying the Sensitivity of Arctic Marine Mammals to Climate-Induced Habitat Change. *Ecological Applications* 18 (2S): S97–S125.

Laidre, Kristin, Mads Peter Heide-Jørgensen, Harry Stern, and Pierre Richard. 2012. Unusual Narwhal Sea Ice Entrapments and Delayed Autumn Freeze-up Trends. *Polar Biology* 35: 149–154.

Lowry, L., K. Laidre, and R. Reeves. 2017. Monodon monoceros. *The IUCN Red List of Threatened Species 2017*: e.T13704A50367651. https://doi.org/10.2305/IUCN.UK.2017-3.RLTS.T13704A50367651.en.

Mantchev, Lisa. 2017. *Someday, Narwhal*. New York: Simon & Schuster.

Marques, Diana Amaral. 2021. *Echolocation Click Rates in Narwhals (Monodon monoceros) as a Conservation Stepping-Stone*. Masters dissertation. Portugal: Universidade de Lisboa.

McLeish, Todd. 2013. *Narwhals: Arctic Whales in a Melting World*. Seattle: University of Washington Press.

Nigg, Joseph. 2013. *Sea Monsters: A Voyage around the World's Most Beguiling Map*. Chicago: University of Chicago Press.
Nixon, Lindsay. 2020. *A Culture of Exploitation: "Reconciliation" and the Institutions of Canadian Art*. Toronto: Yellowhead Institute, Toronto Metropolitan University.
Notz, Dirk, and Jochem Marotzke. 2012. Observations Reveal External Driver for Arctic Sea-Ice Retreat. *Geophysical Research Letters* 39 (L08502): 1–6.
Nweeia, Martin, David Angnatsiak, David Thomas, and Paniloo Guya. 2009. Inuit Qaujimajatuqangit of the Narwhal: Traditional Knowledge Integrated with Tusk Scientific Research. In *Orality in the 21st century: Inuit discourse and practices. Proceedings of the 15th Inuit Studies Conference*, ed. B Collignon and M Therrien. Paris: INALCO.
Pearce, Tristan, James Ford, Ashlee Cunsolo Willox, and Barry Smit. 2015. Inuit Traditional Ecological Knowledge (TEK), Subsistence Hunting and Adaptation to Climate Change in the Canadian Arctic. *Arctic* 68 (2): 233–245.
Pinngortitaleriffik. n.d. Greenland Institute of Natural Resources. https://natur.gl/?lang=en.
Pluskowski, A. 2004. Narwhals or Unicorns? Exotic Animals as Material Culture in Medieval Europe. *European Journal of Archaeology* 7 (3): 291–313.
Pro, Joseph. 2012. *The Adventures of Miki the Narwhal: Miki's Long Journey*. Melbourne: Sky Architects.
Sima, Jesse. 2017. *Not Quite Narwhal*. New York: Simon & Schuster.

CHAPTER 5

"We Will All Be Marine Mammals Soon": Oceanic Intimacy and Extinction from Shakespeare to the Left-to-Die Boat

Steve Mentz

Whales sing songs of extinction from blue depths. From the heady "Save the Whales!" days of environmental activism in the 1970s to today's commercialization of whale-watching tourism, large cetaceans have been powerful symbols that focus our understandings of extinction risk and the biological health of our oceans. These vast swimmers make us feel deeply. Many of us, whether we contact these creatures physically or not, feel touched by cetacean presences. Probably more than any other marine creature, cetaceans catch human emotions. Their massive bulk, alien intelligence, and visible vulnerability have made them potent symbols of oceanic mysteries and changes in our global waters. But these emotionally powerful connections represent a relatively new phase in the cultural history of humans and whales. These creatures have entered the category of "charismatic megafauna," which emphasizes their emotional value for humans. The feelings of intimacy and dislocation that whales generate

S. Mentz (✉)
St John's University, Jamaica, NY, USA
e-mail: mentzs@stjohns.edu

© The Author(s), under exclusive license to Springer Nature Switzerland AG 2024

V. Fibisan, R. Murray (eds.), *Blue Extinction in Literature, Art, and Culture*, Palgrave Studies in Animals and Literature,
https://doi.org/10.1007/978-3-031-69910-8_5

today contrast sharply with early modern understandings of these creatures as symbols of predation and destruction. As whale hunting has given way to whale conservation, a taste of violence continues to follow these vast bodies. In both modes whales are powerful, and in fact Graham Huggan's recent suggestion that whales represent "not just a profusion of allegories but the making of allegory itself" (Huggan 2018, 86) may hold true across the divide between early modern and modern eras. Whales touch human bodies and minds as symbols and warnings, carrying traces of violence and intimacy. Ideas about whales trace a complex history of how we feel about the world's oceans. The threat of extinction may apply only to the modern whales that we love, but traces of vulnerability also appear in stories about threatening premodern cetaceans.

This chapter interrogates the multiple meanings of human-cetacean relations through a rallying cry that I take from the stirring 2020 book *Undrowned: Black Feminist Lessons from Marine Mammals* by Alexis Pauline Gumbs. Gumbs's engagement with cetaceans combines visceral identification with political urgency. She insists on thinking with and about whales and humans together, and this chapter follows her in the effort to connect these mammals and their cultures with humans and our cultures. I use Gumbs's framework to explore a series of representations of the relationships between whales and humans from the early modern period through today. The complexity of these relationships requires multiple critical and historical perspectives. Two competing models for the human-whale relationship emerge: an early modern one based on violence and a modern one based on love. My analysis of these contrasting and entangled modes unfolds across four sections: Encounters, Cultures, Beached Whale Media, and Whales Beyond Water. This admittedly complex structure helps make sense of the changing relationships between humans and whales across historical time and into an age of possible extinctions and ecological catastrophes.

Those of us who are repeated readers (or teachers) of Herman Melville's *Moby-Dick* may have already considered that it will prove impossible to disentangle violence from love. Ahab's killing rage sails on the same ship as Ishmael's worship of the whale's visible presence. All four sections of this chapter draw inspiration from Melville's obsessive intertwining of love and violence. Braiding these strands together, I will by way of conclusion briefly splash into my creative-critical book of poetry, *Sailing without Ahab* (2024), which reconstructs Melville's whale hunting epic by preemptively jettisoning Melville's most famous human creation, the Captain. As my poems suggest, the desire to sail without Ahab, without hunting, and without violence, to live with whales without killing them, proves both tempting and elusive. It may be that we cannot love these creatures without doing them violence, or perhaps even our violence has all along been a form of love. But in the face of the Sixth Great Extinction and Anthropocene climate catastrophes, extreme measures may become necessary soon.

Encounters (Gumbs, Greenwich, and Melville)

No creature, save perhaps the dodo, speaks more directly to fears of anthropogenic extinction than whales. The massive cultural surge in interest in the songs of humpback whales, which were recorded and circulated in the early 1970s, flowed from a collective fear of cetacean extermination. These haunting songs, as Margert Grebowicz observes in her brilliant book *Whale Song*, "were the soundtrack to imagining whale extinction and environmental death" (Grebowicz 2017, 11). In taking up questions of whales and extinction in the context of an inter-species blue humanities, I reach back to earlier times and technologies of whale-human encounters, including the intimate violence of whale hunting. Building on Dan Brayton's observations about the role of early modern cetaceans in "mediating the relationship of humanity with nature" (Brayton 2012a, 115), I consider the dying whale as an emotionally charged symbol of anthropogenic environmental damage.

Alongside images of violence against whales I juxtapose Gumbs's claim that marine mammals can teach the lessons of Black feminism, including radical empathy and connection with vulnerable bodies. Enjoined by Gumbs to become "a marine mammal apprentice" (Gumbs 2020, 3) and to cultivate "undrowning" as a survival practice during our era of sea level rise, I use whale encounters to suggest how these creatures can expand our ecological thinking, inspire oceanic intimacies, and re-order ideas of the human. Like many who have encountered Gumbs's book since its publication during the descent into pandemic winter of November 2020, I have been drawn to her passionate speculations and direct claims of connection to marine mammals. Her assertion of continuity with whales through the Black feminist legacies of suffering—"I am related to all marine mammals," she claims, "I am related to all those in captivity" (Gumbs 2020, 37)—intermingles marine and terrestrial environments. She connects whales to movements for human and environmental liberation:

> I can't help but notice how marine mammals are queer, fierce, protective of each other, complex, shaped by conflict, and struggling to survive the extractive and militarized context our species has imposed on the oceans and ourselves. (Gumbs 2020, 9)

The survival of these "fierce" sometimes black-skinned creatures becomes, for Gumbs, a "celebration" (Gumbs 2020, 67). To humanize whales and connect to their queer conflicts and endurance means also to cetacean-ize ourselves—to posit that "we will all be marine mammals soon" (Gumbs 2020, 117), which entails both preparing ourselves for the waters that will rise and also liberating ourselves into whale-habits of being. Gumbs finds in the same-sex attachments of the common bottlenose dolphin—"Flipper and them" (Gumbs 2020, 117), she names these familiar creatures—inspiration for erotic and personal change. "So thank you for remembering to swim," she says, "despite everything" (Gumbs 2020, 117). Her cascading vision, full even to excess, frames my discussion of human-whale encounters as promise and lure. Few humans, perhaps not even Ishmael himself, seek more from whale bodies than Gumbs does. It may be that her example can help surface a supporting undersong hidden inside the historical master-plot of violence.

Alongside Gumbs's fantasia of queer empathy I present a series of fatal whales, including one of uncertain species that swam up the Thames estuary in June 1658, just past where the Tidal Barrier sits today in Greenwich. The whale's brief and violent London adventure appears in a short prose pamphlet that treats the whale's arrival as one of many "messengers of God, to forewarn us, and fore-arm us, against the dreadfull day of the Lord" (*Londons Wonder* 1658, sig. A3v.). The human reaction to this whale's incursion into urban waters was swift and predictable: "some Water-men thereabouts … instantly got Harping-irons: Spits, Hatchets, Bills, and Axes, & fell a striking the Whale as far as they durst ventur" (*Londons Wonder* 1658, sig. A4). Readers of *Moby-Dick* will perhaps be inured to the bloody scene that follows, but by way of contrast to Gumbs's lyric hymn to connection, I want to quote a bit of the cruelty: "the vehemency of [the whale's] Wounds made her disgorge the water out of her mouth in abundance, that the people there present were amazed to see it" (*Londons Wonder* 1658, sig. A4v). The slaughter becomes a public performance, "and worthy Cittizens went thither in their Coaches" (*Londons Wonder* 1658, sig. A5). The mammal's death requires intimate proximity, until "at last two Ankers belonging, to the fishers, was struck fast into the body of the Whale, which was the cause of her death" (*Londons Wonder* 1658, sig. A5v). It never becomes clear why the pamphlet feminizes the whale, calling the creature "her" throughout. Presumably they determined her sex when they measured and butchered the corpse, though the text only particularizes the size—58 feet long, 12 feet high, 14 feet in breadth, and "two foot long between the eyes" (*Londons Wonder* 1658, sig. A6v). The violence of the kill and fanatical intensity with which the crowd "fell to cutting of her up with axes" (*Londons Wonder* 1658, sig. A5v) points to the powerful emotions and desires called up by the hunt. The anonymous author of the pamphlet insists that the whale in the Thames represents "some signe or token of heavens displeasure" (*Londons Wonder* 1658, sig. A6v). It seems possible that the pamphlet, published during the interregnum when the execution of King Charles I had left the throne vacant, speaks to regicidal impulses and their consequences. But the eager bloodlust of the hunt also reveals primal connections and desires.

Many readers of this chapter, very much including me, return to the doomed hunt for *Moby-Dick* year after year, like salmon to our particular stream. In that context, the blood and cruelty of this seventeenth-century whale-killing pamphlet does not seem unusual. A few years ago, during the height of the Covid-19 pandemic, I steered another semester's boat-crew of greenhorn readers through Melville's epic, this time in the company of Gumbs and other Black feminist writers including (Sharpe 2016; Hartmann 2007; Brand 2001). I can attest, during that strange year and alongside this critical company, that Melville's gore, to which I have become accustomed through decades of exposure, should shock us more than it does. Familiar scenes in which "Stubb Kills a Whale" (Melville 2002, Ch 61) or Ahab soliloquizes "The Dying Whale" (Melville 2002, Ch 116) became, for me and my students during the pandemic spring of 2021, freshly painful. Stubb, in particular, who in the past I have considered a jolly if reckless mariner, assumes upon reflection the hideously cruel face of a leering slave master. As Gumbs's perspective of radical empathy reminds us, he tortures black mammalian flesh. In the context of extinction, however, *Moby-Dick*'s crucial chapter hides itself within Ishmael's meditative mood, following other faux-academic chapters such as "The Fossil Whale" (Melville 2002, Ch 104) and "Measurement of the Whale's Skeleton" (Melville 2002, Ch 103). The meditation on extinction is one of Ishmael's most doubled and self-consuming chapters. He argues himself into opposing positions on the already doubled question of "Does the Whale's Magnitude Diminish? – Will He Perish?" (Melville 2002, Ch

105). The answers, as Ishmael refuses to admit but appears to know, should be yes, and yes. Whales will diminish and will perish. Idealist to the last, Ishmael insists on answering yes but no. The whale diminishes but will never perish. Ishmael's denial of extinction captures his author's rageful Romanticism. Whales as symbols can never die.

Ishmael, Melville, and we who sit at the twenty-first-century end of their long chase, know that whale populations had already started to fall due to pressure by human hunters when Melville published his novel in 1851. Ishmael recognizes the violence exerted by the "almost omniscient look-outs at the mast-heads of the whale-ships" and "the thousand harpoons and lances darted along all continental coasts" (Melville 2002, 352). He knows that in the mid-nineteenth century, when Melville went a-whaling, American whaleships sought their prey in remote Pacific and polar waters, when previously they had found and killed closer to home. But whales to Ishmael, as to Gumbs also, and perhaps still to us, swim as both symbol and creature. The whale must, Ishmael pleads against his own better knowledge and experience, be accounted "immortal in his species, however perishable in his individuality" (Melville 2002, 354). No individual dead whale could threaten the immortal Whale as such.

To read the symbolic Whale through a Black Atlantic lens recasts the blue humanities thinking that many people, including me, have been pursuing over the past decades (see Mentz 2023). What Jonathan Howard calls "a black and blue humanities" (Howard 2022) aims to entangle the stories of suffering and dislocation central to Gumbs's engagement with marine mammals with a rich palette of liberationist thinking. To connect the Black Atlantic to the blue humanities may enable a reimagination of cetacean histories, including the history of their brushes with extinction, violence, and trans-species love.

Culture (Shakespeare, Whitehead, and Cox)

For William Shakespeare in early modern England, whales assumed many guises, but their dominant meaning framed them as violent predators. Whales as described in Shakespeare's plays were mostly aggressive monsters. If these creatures, the largest and most symbolically capacious mammals in the sea, were themselves aggressive, human violence toward them could be imagined as self-defense. As Dan Brayton has demonstrated, a series of symbolic connections link "princes, whales, and dolphins" (Brayton 2012b, 59) as representations of power. Brayton further suggests that the whale is a distorting mirror, being at the same time "an emblem of epistemological uncertainty and … strange kinship" (Brayton 2012b, 59) between humans and the ocean. Their superabundance of mammalian flesh and red blood suggests that whales' "greatness is indistinguishable from that which is gross or base" (Brayton 2012b, 61). In *Pericles*, cetacean power reveals itself in the storm-attracting porpoise, who is described by a fisherman as "half fish, half flesh" (Shakespeare 2004, 2.1.25; see also Mentz 2012). Unseen marine mammals function in this play primarily as cultural symbols. "I can compare our rich misers," says the Fisherman, "to nothing so fitly as to a whale: 'a plays and tumbles, driving the poor fry before him, and at last devours them all'" (Shakespeare 2004, 2.1.29–31). This greedy whale comes to represent depth and opacity when Pericles later refers to Thaisa's body descending to where "the belching whale / And humming water must o'erwhelm thy corpse" (Shakespeare 2004, 3.1.62–63; see also Duckert 2019 and Publicover 2002). Unlike the whale in the Thames who wanders into shallow water, these symbolic whales represent predatory landowners or the unseen profundity of death. With similar violence, in *Troilus and Cressida* Hector in his rage makes the Greek soldiers appear "like scaled schools / Before the belching whale" (Shakespeare 1998, 5.5.22–23). Even outside the violent worlds of tragedy and epic, whales represent predatory power, as in Paroles's misogynist comment in *All's Well that Ends Well* that his friend the Count is "a whale to virginity, and devours up all the fry it finds" (Shakespeare 2019, 4.3.216–17). Whales, in this dominant metaphor, are the biggest predators in the sea.

Neither Paroles nor the fishermen in *Pericles* realized that the whales they watched were also watching them. Perhaps the most suggestive new scientific research on whale behavior has emerged in the work of marine biologists Hal Whitehead and Luke Rendell, who argue that many whale populations possess an ancient, flexible, and inheritable culture. Their 2015 book *The Cultural Lives of Whales and Dolphins* explores the musical culture of humpback whales, about which they note that it "dominates the acoustic environment of the ocean, as it has for millions of years" (Rendell and Whitehead 2014, 2). The two biologists further contrast "human cultural supremacy over the surface of the earth" as being by contrast "recent and not quite complete" (Rendell and Whitehead 2014, 2). More recent research has confirmed that sperm whales in the nineteenth century responded to the flurry of harpoons, including those cast by Melville's historical *Acushnet* and fictional *Pequod*, by devising and spreading habits of defense against human hunters that decreased the success of whalers, at least until the development of diesel engines and the explosive harpoon. Philip Hoare, writing about this research in *The Guardian* in March 2021, paraphrases a Royal Society paper by Rendell and Whitehead that demonstrates that sperm whales learned from their early contact with human hunters. "As their culture made fatal first contact with ours," Hoare writes, "they learned quickly from their mistakes" (Hoare 2021b). Industrial whaling techniques would overwhelm these cultural adaptations (see Demuth 2019). As Michael Moore notes in his powerful 2022 book *We Are All Whalers,* current whale populations, which rebounded in the late twentieth century after the substantial end of commercial whaling, are currently falling again due to entanglements in fishing gear, mostly from lobster pots, and also collisions, mostly with commercial shipping vessels. Moore, who early in his career did research with commercial whalers in the Southern Ocean as well as sailing on a research vessel with Hal Whitehead, emphasizes that all of us who eat seafood or participate in the global economy share blame (Moore 2022).

Many contemporary authors have engaged with whales as symbols, muses, and inspirations, but few write as eloquently and lovingly about whales in literary and human terms as Philip Hoare. In a trilogy of semi-autobiographies, Hoare narrates his life story in dialogue with Melville's epic and his fascination with cetaceans (Hoare 2008, 2013, 2017). Most recently, *Albert and the Whale* (2021) enmeshes Hoare's own distinctive autobiographical voice with the career of Renaissance artist Albrecht Dürer. An enigmatic episode describes Dürer's eager journey to the coast of Zeeland in 1519 to see stranded whales on the beach. By the time the artist reached the coast, the whales had washed back out to sea. Hoare describes this missed viewing as a turning point in Dürer's artistic career. To Hoare, Dürer clearly was cetacean-mad, just like Hoare himself. "Had Dürer seen even one whale," Hoare writes, "his art would have preempted Melville's" (Hoare 2021a, 70). The beached whales loom in these artists' imaginations, "witnesses in the meaning of the word martyr, auguries of their own extinction" (Hoare 2021a, 70). These unseen whales represent an extinction of art, trapped on the beach, "crushed by their own weight" (Hoare 2021a, 70). Whales on a beach, out of their element—perhaps, now that fewer harpoons but more toxins and noise pollution fill their seas, extinction will find them driven to shore rather than swimming at sea.

The violence of the human-whale relationship, as chronicled by Melville among others, asserts a bloody master plot. Writers like Hoare, scientists like Whitehead, and animal rights activists have been proposing alternative storylines since the late twentieth century. One of the most poignant narrative inversions of the hunt for the White Whale appears in Lynne Cox's autobiographical *Grayson* (2006), which describes her swimming with a whale calf off southern California when she was a teenager, training for open water competitions. She encounters a baby gray whale, "between three and four months old" (Cox 2006, 46), who has lost track of his mother. In a precise inversion of the whalemen who greet cetaceans with harpoons, Cox determines to swim with the calf until his mother returns. "Okay, let's go for a swim, baby whale" (Cox 2006, 54), she says, and enters into a multispecies entanglement in chilly Pacific waters. The story ends happily, but only because Cox, who would become the world's most celebrated cold-water endurance swimmer (see Cox 2005), was able to survive in the whale's world for several hours. Few humans can swim with the young Lynne Cox, to say nothing of the baby whale she names Grayson. When at last the massive mother, after she has been reunited with her calf, emits "a series of sounds, high sounds and low ones and probably so many more at frequencies that were too low for any of us to hear" (Cox 2006, 136), the full communication excludes human listeners. But Cox continues to wonder about the life of that young whale, as she continues to swim off the California coast. Grayson, if he still swims, represents a rare case of a whale whose encounter with humans was lifesaving, not violent. Cox's novella-sized memoir may not fully invert *Moby-Dick*'s gargantuan testimony to the human urge to exterminate. But Grayson reminds us that there are other stories about humans and whales that can be told.

Beached Whale Media (Shakespeare, Peters, and Giggs)

Perhaps the most quotable whale in Shakespeare never touches the water, though its body consists of (possibly imaginary) water vapor. In *Hamlet*, the prince badgers Polonius into agreeing that the cloud they may not even see appears "very like a whale" (Shakespeare 2006, 3.2.373). The episode, which is staged indoors and at night, captures Hamlet's antic aggression. These are only whale-clouds of the mind. But the shifting typologies of cloud-figures through which Hamlet drags the old counselor, from camel to weasel to whale, recall the excessive generic hybridizations with which Polonius had greeted the newly arrived players. Imagined camel-weasel-whales are not exactly the "comical-tragical-historical-pastoral" plays (Shakespeare 2006, 2.2, Q1) that the old man celebrates, but the processes through which the cloud-animals and the genres transform echo each other. (For a muted defense of Polonius as genre theorist that I hope someday to make more full-throated, see Mentz 2006, 29.) Arriving at the whale as the final figure in the cloud parade suggests that cetaceans represent a symbolic end point, an accumulation of flesh, form, water, and potency beyond which even the prince of Denmark cannot pass.

The media critic John Durham Peters also thinks about whales in dialogue with clouds and media forms. For Peters, the divergence between the brains of whales and humans emerges from their different environments: "Cetaceans are born in the ocean medium, but humans cannot live there without craft" (Peters 2015, 56). The distinctive feature of the "marine habitat is that it filters out light and enhances sound conductivity" (Peters 2015, 61), which Peters suggests inverts many of humanity's most basic habits of perception and morality. "What would an ethics be like," he wonders, "that did not take the face as a claim to individual dignity and the voice as a representation of will and choice? Perhaps cetacean empathy is performed corporeally and gymnastically rather than facially, or perhaps cetaceans are exquisitely sensitive readers of the microtones in each other's sound productions" (Peters 2015, 68). Connecting the heavy wet media of the ocean with the thin dry air in which clouds float, Peters further suggests, with reference to Hamlet's clouds, that these airy forms are "the thing par excellence without inherent meaning" (Peters 2015, 254). Whales represent perfect but nonvisual communication while clouds are transient shapes. Both represent themselves in watery forms.

Hamlet's clouds and Peters's media theory create abstractions, whales without bodies. These wispy visions occlude, at least temporarily, human violence toward cetaceans. But when we turn our gaze to real whales who have been stranded on real beaches, bodies is all that they are. Lest we think that whale suffering and death have been consigned to the bad old days, a recent stranding attracted the attention of the contemporary Australian writer Rebecca Giggs. She opens her 2020 book *Fathoms* with a slow and painful description of its slow and painful death:

> By sunrise, a part of the whale that ought not to be outside of it was outside of it. A digestive organ, frilled and bluish in the foam. The whale's billiard-ball eyes tumbled in its head, and its breathing sounded labored. … Seagulls flew down to peck avian hieroglyphs in the whale's back, their inscriptions legible to yet more skyward gulls that dove to elaborate the wounds. At every nip the whale flinched, still intensely alive. (Giggs 2020, 3)

The horror of that pain, the animal girth suffocating itself through its exposure to a world without buoyancy, transforms the whale from symbol of power to representation of flesh. For Giggs, the whale becomes a world, ecological and cultural. In the moment, in the air, fragile, and in pain, its dying body occupies the place that pain takes all bodies. Away from communication, alone in respiratory failure, isolated out of its element, the stranded whale breathes, until it stops.

My own single encounter with a beached whale occurred when I was a college student in 1985, visiting Oregon and the west coast of the United States for the first time. The stranded cetacean I saw had progressed farther down the path of biological inevitability than the body Giggs saw in Australia. I heard no breathing and saw no flinching. The poor creature had probably stranded a few days before I arrived. The flesh was barely recognizable as whale when I walked past on my way toward the water. Its skin was suppurating, and its flesh had blown into a thick pink paste. Nothing I have encountered before or since prepared me for the intensity of the smell. Sulphur, rotten eggs, over-ripe meat, perhaps the sharp acrid pinch of cat urine on an old rug—all these odors hint at how whale-smell overwhelms, a tidal smell, physical, pressing hard and intensely into your body, like a cork just released from watery depths. I wanted to look, to bear witness, and to recognize what could still be recognized, but I could barely make myself walk close. The smooth contours of whale-flesh sagged into semi-gelatinous liquidity as the stench held me at bay. I wonder if I will ever get closer to a whale. The smell seemed a physical force, pushing me away and working my emotions. The whale-body that for Prince Hamlet found expression in imaginary cloud-shapes confronted me as an invisible but physical force. I can smell it still.

Whales Beyond Water (Shakespeare, Brin, and Bergvall)

Shakespeare also presents whales on the beach, but for him the shore is a symbolic location not a space of visceral encounter. In these cases, the predatory force of the great animal gets inverted. As Mistress Ford mocks Falstaff in *Merry Wives of Windsor*, "What tempest, I trow, threw this whale, with so many tuns of oil in his belly, ashore at Windsor?" (Shakespeare 2000, 2.1.56–58). Once Falstaff has been removed from the elite political world of the *King Henry IV* plays, his out of place-ness becomes simply comic. The fat knight becomes just a pile of beached oily flesh. But whales on the beach also have more sinister political valences. King Henry IV uses the metaphor to paint his son the prince's moodiness as self-destructive: "Till that his passions, like a whale on the ground, / Confound themselves with working" (Shakespeare 2016, 4.3.40–41). The whale's vain struggle to survive in an alien environment becomes, for the king, a necessary if brutal stage in the education of his ne'er do-well heir. Henry's metaphor imagines the prince "working" like a stranded whale, instrumentalizing the agony that these marine creatures endure when out of their native element.

The violence in King Henry's image of the working beached whale has given way more recently to New Age visions of cetaceans as spiritual guides. As analyzed by Mette Bryld and Nina Lykke in their book *Cosmodolphins: Feminist Cultural Studies of Technology, Animals, and the Sacred* (2000), dolphins became in the late twentieth century potent symbols of New Age transcendence and transformation. They observe that for many writers and thinkers "the dolphin and the whale will allegedly guide us to insight into the 'true and sacred' pleasures of a simple life in harmony with the natural environment" (Bryld and Lykke 2000, 3). These two feminist scholars further note that dolphins and whales have become iconic figures not just in New Age spirituality but also in the different futuristic visions of science and science fiction. (One of Bryld and Lykke's primary examples is Joan Ocean, who claims that the supernatural nature of dolphins as visitors from the cosmic beyond was revealed to her in an acoustic image transmitted by the animals themselves, with whom she swam in Hawaii. Ocean's most startling claim is that dolphins are recent arrivals on earth from a higher "Universal reality." See Ocean 1997, 163.) Bryld and Lykke note that "dolphins entered the American imagination around 1960" (Bryld and Lykke 2000, 183), largely through the idiosyncratic science of John C. Lilly, whose book arguing for the imminent possibility of dolphin-human communication, *Man and Dolphin*, appeared in 1961. Bryld and Lykke emphasize that for scientists such as Lilly, and for sci-fi rationalists such as David Brin, author of *Startide Rising* (1983), cetaceans can assume cyborg identities, following an "uplift" procedure in which humans mechanically increase their companion species' linguistic and other mental capacities. Bryld and Lykke note that *Startide Rising*, with its spaceship operated by a mostly dolphin crew, ultimately endorses a humanist and human-centric position, by contrast with the cyborg feminism advocated by Donna Haraway among others since the 1980s (Bryld and Lykke 2000, 69). From a whale-centric point of view, however, it seems notable that the dolphins in Brin's novel repeatedly describe their humpback whale kin as being the great poets of the ocean, whose haiku-like verses comprise the "Whale Dream" that all dolphins idolize. This poetic world serves the Uplifted starship crew of "fins" as a kind of oceanic scripture:

> As the humpback—
> older sibling
> sings songs to the –
> Serious fishes (Brin 1983)

Even in Brin's technologically driven universe, the songs of humpback whales represent spiritual connections. In his sci-fi world, as also in New Age fantasy, whale spirituality seems a displacement of historical violence, a shifting of the point of humanity's killing spear in a non-bodily direction.

An even more intimate human-whale entanglement appears in a poetic scenario in which whales and humans merge in flesh as well as spirit. In an episode that seems almost to literalize Gumbs's utopian hopes and the metaphysical musings of Ishmael on the masthead, the French-Norwegian writer and performance artist Caroline Bergvall reimagines whales in her rendition of the anonymous Old English poem, "The Seafarer." Her experimental book *Drift* (2014) mashes together the early medieval elegy with the story of the Left-to-Die Boat. This infamous boat was an overloaded vessel carrying 72 African refugees from Tripoli toward Italy in March 2011. After floating in the Mediterranean for 15 days, in frequent view of multiple NATO ships and aircraft, the boat washed back up on Libya's coast with only 11 survivors, two of whom subsequently died. Alongside this human and political maritime tragedy, Bergvall juxtaposes the anonymous voice of the seafarer, isolated on the ocean, exiled from warmth, home, and humanity.

In Old English and other Nordic languages, the ocean is the "whale's road," the "hronrade" in the tenth line of *Beowulf* (*Beowulf* 1950, 1). In Bergvall's hybrid poetic language the corresponding space is "hafville," which she describes as "sea wilderness, sea wildering" (Bergvall 2014, 3). Trailed by an anonymizing refrain, "Blow wind blow, anon am I" (Bergvall 2014, 53), her seafarer plies northern emptiness. "Whaling girls" appear briefly (Bergvall 2014, 53). The seafarer defends his lonely craft against "the kraken the belching hafgufa the / heather-backed whale almighty" (Bergvall 2014, 52). At the end of the sequence of songs, the poet invites the whale into bodily embrace:

> Let
> the whales have me todaeg Let me fold into the
> fattened tissue of these large gentle ocean giants
> Eagles die Kinsmen die You yourself will die One
> thing never dies the fate of the honoured dead (Bergvall 2014, 58)

The seafarer enters whale's flesh and whale's road, his body merges with fat and stink. The poem sings a vision of personal extinction, in a way, the extermination of the heroic individual, with a partial gesture toward communal after-living through song. The seafarer becomes a marine mammal, not in Gumbs's sense or even Hoare's, but in a visceral, painful, self-extinguishing way. As twentieth- and twenty-first-century human technologies threaten the modern descendants of these gentle giants, perhaps our own fattened tissue is also at risk.

Conclusion: The Whale in Labrador Pond

What if the whale came to us?

I close this speculative chapter with a poem that I wrote for *Sailing without Ahab*, a book that assembles a chapter-by-chapter response in 138 poems to Melville's epic voyage. I begin the trip by jettisoning the Captain in name and symbol. This poetic voyage refuses to sail with authoritarianism, domination, and tyranny. For the conclusion of this chapter, I select a poem that relocates a whale from out of the deep sea to a shallow lake in rural Maine, where I spent some summer vacations with my family when my children were small. In this poem, the whale represents a connection with oceanic space, but the creature is simply too vast for the inland watery body around which we spent a happy summer week. Like so many of the texts I've explored in this chapter, the poem links whales and humans while also showing that full intimacy between us is impossible in practice. The poem imagines, but cannot quite present, a world big enough for all. The "I" that is both myself and whale-idealist Ishmael dreams of a utopian escape from the threat of extinction.

> Out of Place
>
> There's a whale in Labrador Pond.
> It's too big.
> In a freshwater pond only about a half-mile wide,
> Each edge rounded off by lily pads and marsh,

Fed by a trickle through a rusty under-dirt-road culvert,
There's no place for a whale.
He came here with me.
I. finds no place else for him.
So in he splashes, fin-full and overflowing,
Displacing with massive girth
Water that surges onto the porch to soak my feet where I. sits
With my family playing Clue.
It's hard to know who did it.
The whale's swimming stirs up green froth.
Its body rips algae-fronds from the silty bottom.
  Air bubbles ascend, balloon, and pop.

In the morning I. wakes to devastation.
  I. brews coffee and hot chocolate and bakes biscuits while sleepers dream
  Of a bigger place (Mentz 2024, 16)

In this poetic fantasy the whale overflows a snug family space. The small house and small lake cannot contain cetacean girth. The intimacy that the whale promises founders on the limits of size and space. Chaos ensues. Our little world becomes "devastation." I find no place for this vast and beautiful creature. But that is what dreaming is for.

References

Anonymous. 1658. *Londons Wonder*. London. Accessed via EEBO. 11 May 2021.

———. 1950. *Beowulf.* Fr. Klaeber, ed. 3rd ed. Lexington: D.C. Heath and Company.

Bergvall, Caroline. 2014. *Drift*. Brooklyn: Nightboat Books.

Brand, Dionne. 2001. *A Map to the Door of No Return: Notes to Belonging*. Toronto: Doubleday Canada.

Brayton, Dan. 2012a. Royal Fish: Shakespeare's Princely Whales. In *The Indistinct Human in the English Renaissance*, ed. Jean E. Feerick and Vin Nardizzi, 47–65. London: Palgrave.

———. 2012b. *Shakespeare's Ocean: An Ecocritical Exploration*. Charlottesville: University of Virginia Press.

Brin, David. 1983. *Startide Rising*. New York: Ballentine.

Bryld, Mette, and Nina Lykke. 2000. *Cosmodolphins: Feminist Cultural Studies of Technology, Animals, and the Sacred*. London: Zed Books.

Cox, Lynne. 2005. *Swimming to Antarctica: Tales of a Long-Distance Swimmer*. New York: Mariner Books.

———. 2006. *Grayson*. New York: Harvest.

Demuth, Bathsheba. 2019. *Floating Coast: An Environmental History of the Bering Strait*. New York: W.W. Norton.
Duckert, Lowell. 2019. *Pericles*'s Deep Ecology. *Studies in English Literature* 59: 367–381.
Giggs, Rebecca. 2020. *Fathoms: The World in the Whale*. New York: Simon & Schuster.
Grebowicz, Margaret. 2017. *Whale Song*. London and New York: Bloomsbury.
Gumbs, Alexis Pauline. 2020. *Undrowned: Black Feminist Lessons from Marine Mammals*. Chico: AK Press.
Hartman, Saidiya. 2007. *Lose Your Mother: A Journey Along the Atlantic Slave Route*. New York: Farrar, Straus, and Giroux.
Hoare, Philip. 2008. *Leviathan, or the Whale*. London: Fourth Estate.
———. 2013. *The Sea Inside*. London: Fourth Estate.
———. 2017. *risingtidefallingstar: In Search of the Soul of the Sea*. Chicago: University of Chicago Press.
———. 2021a. *Albert and the Whale: Albrecht Dürer and How Art Imagines Our World*. New York and London: Pegasus.
———. 2021b. Sperm Whales in Nineteenth-Century Shared Ship Attack Information. *The Guardian*. https://www.theguardian.com/environment/2021/mar/17/sperm-whales-in-19th-century-shared-ship-attack-information. Accessed 29 February 2024.
Howard, Jonathan. 2022. Swim Your Ground: Toward a Black and Blue Humanities. *Atlantic Studies* 20: 308–330. https://doi.org/10.1080/14788810.2021.2015944.
Huggan, Graham. 2018. *Colonialism, Culture, Whales: The Cetacean Quartet*. London: Bloomsbury.
Lilly, John C. 1961. *Man and Dolphin*. New York: Doubleday.
Melville, Herman. 2002. *Moby-Dick, or The Whale*. Ed. Hershel Parker and Harrison Hayford. New York: Norton.
Mentz, Steve. 2006. *Romance for Sale in Early Modern England*. London: Routledge.
———. 2012. Half-Fish, Half-Flesh: Dolphins, Humans, and the Early Modern Ocean. In *The Indistinct Human in the English Renaissance*, ed. Jean Feerick and Vin Nardizzi, 47–65. London: Palgrave.
———. 2023. *An Introduction to the Blue Humanities*. London: Routledge.
———. 2024. *Sailing without Ahab: An Eco-Poetic Voyage*. New York: Fordham University Press.
Moore, Michael. 2022. *We Are All Whalers: The Plight of Whales and Our Responsibility*. Chicago: University of Chicago Press.
Ocean, Joan. 1997. *Dolphins Into the Future*. Kailua: Dolphin Connection Books.
Peters, John Durham. 2015. *The Marvelous Clouds: Toward a Philosophy of Elemental Media*. Chicago: University of Chicago Press.

Publicover, Laurence. 2002. *Pericles*'s Humming Waters: Nonhuman Agency, Textual Criticism, and the Practice of Material Ecocriticism. *Shakespeare Quarterly* 73: 280–302.
Rendell, Luke, and Hal Whitehead. 2014. *The Cultural Lives of Whales and Dolphins*. Chicago: University of Chicago Press.
Shakespeare, William. 1998. *Troilus and Cressida*. Ed. David Bevington. London: Arden 2.
———. 2000. *The Merry Wives of Windsor*. Ed. Giorgio Melichiori. London: Arden 3.
———. 2004. *Pericles*. Ed. Suzanne Gosset. London: Arden 3.
———. 2006. *Hamlet*. Ed. Ann Thompson and Neil Taylor. London: Arden 3.
———. 2016. *King Henry IV, Part 2*. Ed. James C. Bulman. London: Arden 3.
———. 2019. *All's Well That Ends Well*. Ed. Suzanne Gossett and Helen Wilcox. London: Arden 3.
Sharpe, Christina. 2016. *In the Wake: On Blackness and Being*. Durham: Duke University Press.

CHAPTER 6

Held Together: Learning Attachment from Oysters

Megan Hayes

Oyster Story

Attachment is a pivotal moment in the life of an oyster. It marks the creature's transformation from one who drifts, to one who will remain moored. Prior to attachment, an oyster lives as a free-swimming larva, roaming the currents of the brackish waters of their birthplace. At around two and a half weeks of age, the larval oyster develops a foot with which it wanders the bottom of the water column in preparation for permanent attachment to a hard surface (Nelson 1924). After choosing their point of attachment, the baby oyster secretes a glue, cementing their body around a single point of contact. Glands in their tiny foot release a protein-rich bioadhesive (uniquely cement-like when compared to the adhesives of other marine organisms like mussels and barnacles (Burkett et al. 2010)) allowing them to hold on amidst the force of waves, currents, and flows (Foulon et al. 2018). This vital attachment allows the oyster to undergo a necessary and massive

M. Hayes (✉)
University of Oregon, Eugene, OR, USA
e-mail: mhayes8@uoregon.edu

© The Author(s), under exclusive license to Springer Nature Switzerland AG 2024

V. Fibisan, R. Murray (eds.), *Blue Extinction in Literature, Art, and Culture*, Palgrave Studies in Animals and Literature,
https://doi.org/10.1007/978-3-031-69910-8_6

metamorphosis of their internal anatomy, spurring their transformation into 'spat.' Once spat, the oyster will begin to grow very quickly, gobbling algae out of the water column and blooming into the fullness of their life.

Oysters draw many others in. Impressively generous ecological inhabitants, an individual oyster can filter up to fifty gallons of water per day (Wilbon et al. 2011), and so, working together, oysters are able to dramatically improve water quality. The nooks and crannies of oyster reefs have vastly greater surface area than that of the smoother ocean floor and, as such, are home to many forms of marine life (Dame 2012, 128). Not only do oysters lure other kinds of creatures, but also one another: they love to be together. While they can find stable ground for attachment on a myriad of solid things, the larval oyster's preferred site for settling in is an existing oyster shell. It is thanks to this preference for attaching to one another that the collective form and strength of an oyster reef is sustained. Nestling beside and atop one another, oysters can create enormous structures out of calcium carbonate, cemented together by generations of bivalves. Each oyster secretes layer inside layer of shell, using minerals pulled from their watery surround (and sequestering carbon in the process). Accumulated together, these shells form a spatially complex substrate and mosaic-like topography that oysters, and a multitude of other creatures, depend on for food and habitat (Lenihan 1999). For an animal who will only ever know one place for the rest of their life (aquacultured oysters notwithstanding), they want to choose a comfy spot. How fitting it is, then, that a reef of oysters resting upon each other, growing together, is also called a bed.

Compared to the life of an individual oyster, an oyster bed endures. Beds are fortified by empty shells undergoing the slow taphonomic process of passing from the biosphere to the lithosphere, which is to say they are living shorelines, rich with death. These beds become environments and planetary embankments, holding ecologies—worlds—together. Despite this, oyster beds are the most threatened marine habitat on the planet, and 85% of them have been lost worldwide over the past 150 years (Beck 2011, 108). The pollution, over-harvest, and development that have driven this (and many more forms of) loss are intimately entangled with extractive relationships bearing no reciprocity with the living world. While native oyster beds were once the ecological and economic heart of many estuaries, they now teeter on the edge of functional extinction worldwide. Of the remaining wild harvest of native oysters, three quarters are drawn from only five ecoregions in North America, but the condition of most of these reefs is "poor at best" (Beck 2011, 107).

The lion's share of this century and a half of loss is part of the increasingly familiar story of entangled colonial and capitalist commercial exploitation of land and sea—spaces that are not ultimately discrete (Guerin 2019). Damaging harvest practices have scraped away at the reefs, sometimes tearing them down to the ocean floor (harmful for oysters who love to be piled high up in the water column); erosion and runoff from industry and coastal overdevelopment have made water uninhabitable through toxicity and smothering sedimentation, as have the effects of upstream activities like altered river flows, dams, and poorly managed agriculture and water quality (Beck 2011; Kirby 2004). As oysters decline in health and numbers, their special ability to filter water is diminished, resulting in poorer water quality—an incredibly difficult cycle to reverse.

Attachment

Attachments hold things in place. This is true of affective attachments as well as biomineral ones. They are points of connection *sustained* through space and time, and *that sustain us* through space and time. Many connotations of attachment in a negative valence (she was too attached to 'this' or 'that') affirm an ideal liberal subject who strives to relinquish dependency. But attachments are necessary for life. One might sing the praises of non-attachment while being quite attached, in fact, to their own life. As Tim Ingold writes, "the foetal blob of the mammalian infant, attached to the interiority of the womb by the line of the umbilical cord, is expelled at birth only to reattach itself externally by clinging digitally to the maternal body" (Ingold 2015, 5). Attachments shore up worlds and cultivate futures. They are critical for the ongoingness and reproduction of life—and so come to feel especially important in times of crisis, when there is a shared sense that life's ongoingness and reproduction are starting to glitch (Berlant 2016).

Attachments are orientations, because they pull in certain directions. The ongoing labor of attachments gives shape to bodies and their movements. Through their co-constitutive nature, the attachments tethering individual bodies and subjectivities confuse formerly clear boundaries, such that the possibility of a sovereign being begins to falter. Attachment focalizes the world otherwise to the limits of the 'I,' blurring the edges of inside and out. The accumulation of attachments can stretch one out thin or bolster one snuggly, depending on their character, because attachments are *structures* of relationality which can be textured in a myriad of ways.

Lauren Berlant describes optimistic attachment as an investment in continuity—one's own, or that of a particular world. In our contemporary moment, the objects to which we have become attached are increasingly unstable, "the encounter with which is the condition under which the historical present appears to us: first as a sensory crisis or disturbance, a threat to the reproduction of life in which loosely knotted subjects proceed in incoherent responses to a world that is at once in disarray, and yet not politically unpredictable enough" (Berlant 2012). Not unpredictable enough, because the forces continuing to squeeze non-normative lives out of existence are still yet to be ruptured. Yet disarray abounds, and the conditions of the future cannot be promised, setting the scene for chaotic attachments. Attachments that can become cruel, when they are an optimistic tether to that which is an obstacle to true flourishing (Berlant 2011).

Societies and ecologies are constellations of attachments. The relations of beings to one another and their surroundings depend on attachments that manifest as habits and repetitions, holding things together just so. Describing the theoretical veer toward 'assemblage' that has taken place in both ecology and anthropology, Elaine Gan and Anna Tsing elucidate how the concept makes sense of how things hold without assuming the social solidarity tethered to notions of "community" (2018, 103). Social and ecological assemblages are figured as looser groupings than communities, and "with growing evidence of symbiosis and interdependence from ecologists and evolutionary biologists … theorists have reimagined the assemblage not as a static set of autonomous elements but as a dynamic process of 'becoming with'" (Gan and Tsing 2018, 103). Even if 'becoming with' unfolds amongst a loose grouping, it is a deeply ethical proposition of entanglement and responsibility (Wright 2014). The figure of the knot has emerged as a way to envision the process and cohesion of becoming with, and it is within this conceptual scaffolding and scholarly lineage that the gravity of attachment presses a little deeper.

Knots are indeed one kind of attachment. A knot, across its various definitions, is tension held in form. Knots keep things in place on the strength of tension. To feel oneself stretched by tension to the point of strain marks the ways in which the body is pressed and pushed on, tugged at, weighed down by force it must endure, the topography of which can be read in a body through its knots. Within the body, knots are the sites of ongoing tension held in the fibers of the muscle or connective tissue, becoming spectral spaces of gripping where the body has been butting up against the world. Repeatedly straining physical actions will pull a body into knots, but then so too will the force of affective weight, from which knots might also start to appear in the belly. These knots will form in any body growing through the endurance of strain, like the knobbly part of a tree branch or a shell. These knots that are the consequence of tension can be read differently than knots anticipating it, as knots wishing to keep something in place, such as those holding an oyster farmer's boat secure at the dock before a storm. But in both cases, the knot is an introduction of fixity into a scene of movement, regardless of whether that fixity gives rise to painful stuckness or stability. If, as Donna Haraway has written, the whole world is "a knot in motion" (2003, 6), then an oyster tense under pressure has to yank the world a little this way or that. Knots and/as attachments are clumpy and uneven. The manner in which they hold certain beings and ways of being in place, some more strained than others, is determined by the structuring of power that differentially allows bodies to endure. Attachment operates across multiple registers when we think about worlds knotted together by oysters. These special bivalves cultivate diversely populated ecological worlds like those mentioned earlier, but also worlds of repair (through restoration projects), salvation (by way of their capacity for water filtration, wave attenuation, and shoreline embankment), reproduction (embankment of the present state of things), and something resembling hope (an ambivalent optimism, perhaps).

The World Bound to Itself

Much work is being done to restore oyster populations. Initiatives implemented by organizations like the National Oceanic and Atmospheric Administration (NOAA), the Billion Oyster Project (BOP), or even the multinational oil and gas company Shell, can entail reviving those on the brink of death, re-enlivening spectral populations rendered extinct or thereabouts, or introducing oysters into spaces they have never previously set foot. These projects are sometimes motivated by the desire to hold onto species before they disappear entirely, although in some hands they can be predicated on narratives of return to some ideal (and ahistorical) environment. Often, oyster-oriented projects are undertaken in service of something a little askance to the oyster itself. Infrastructural projects using oyster beds as embankments have their gaze cast toward impending futures. These endeavors are pursued as a means of protecting human populations from the worst effects of environmental degradation and climate crisis: work that does not necessarily seek to change unsustainable modes of life, but rather buttresses them. Borrowing from Berlant, I want to think about structure as "that which organizes transformation," and infrastructure as "that which binds us to the world in movement and keeps the world practically bound to itself" (2016, 394). It is with these definitions in mind that I will turn to some specific oyster projects and consider how their attachments structure worlds.

BOP, based in New York City, is a particularly notable example of a large-scale oyster restoration project. On account of the thickly knotted metropolitan interface of its brackish waters, BOP is one of the most well-known and well-branded oyster restoration projects in the world. On their website, BOP describes the propensity of oysters to grow atop one another as the creation of "a hardy infrastructure for a lively underwater city of marine wildlife." This "infrastructure" seeks to return oysters to the once abundant New York Harbor as a means of drawing in further marine life, cleaning the harbor waters, and building socio-ecological worlds around the Eastern Oyster. New York Harbor was once flush with oyster beds, and the Lenape people lived with, tended to, and ate the oysters of what is now known as New York City for thousands of years before Europeans acquired a taste for them (Kurlansky 2006; Guerin 2019). The thirst for these oysters in early colonial Manhattan—a culture necessarily built on values of extraction and exploitation—exhausted populations that had been carefully stewarded for millennia.

Today's New York Harbor is a very different place. Its waters have been impacted by shipping, urban development, and pollution, and to this day it is flushed with untreated sewage and city street detritus when it rains, thanks to the city's Combined Sewage Overflow (CSO) system. Into this perpetual cycle of deferred waste, the oyster is optimistically pulled. As filter feeders, oysters have the magical power of cleaning water just by eating. The fifty gallons of water an oyster can filter every day is the basis of BOP's arithmetic: the one billion oysters they plan to 'install' in New York Harbor by 2035 will, in theory, filter the entire body of water every three days. On account of the quantity of pollutants these harbor oysters are made to ingest with every meal, they are strictly categorized as infrastructure—not food. But in addition to their water-cleaning efforts, these oysters are also recruited to build cultures of work, education, and recreation that create an explicitly 'environmental' point of connection for the human population of one of the planet's densest urban centers.

BOP is also working in collaboration with Kate Orff and SCAPE Landscape Architecture Studios to install oysters on and around the Living

Breakwaters project—an exemplary model of coastal 'green infrastructure' which seeks to mimic the reef systems that once hemmed southern Staten Island. In the wake of Hurricane Sandy in 2012, the U.S. Department of Housing and Urban Development held the Rebuild by Design Competition, in search of "cutting edge" design ideas for coastal resilience (New York State Homes and Community Renewal). The historic reef systems that SCAPE's design seeks to remember had been a source of food, but also buffered the city by reducing flooding, mitigating erosion, and softening the intensity of large waves. By reclaiming those oyster attachments, Living Breakwaters seeks to build protective knots of attachment around New York City's coastline in the face of escalating climate crisis. Emily Eliza Scott has detailed the function of SCAPE's infrastructure "as a substrate, or scaffold: not solid ground, but a platform upon which multiple human and nonhuman actors are set into motion and interactions are expected to evolve over time" (Scott 2016, 187). Living Breakwaters is built on, and cultivates, "networks, processes, flexibility, and porosity, or 'soft' edges." It is a worldbuilding project that creates connective tissues out of oysters: entanglements of attachment antithetical to modern infrastructural paradigms that seek to control discrete objects.

So-called green infrastructures imagine the world bound to itself through a skin of mediation and management. Existing or exhausted natural processes are employed to keep (some) worlds moving, often with the assertion that ecosystems were always already infrastructures—the definition of 'natural' becoming confused as a result. For Sara Nelson and Patrick Bigger, "infrastructural nature" names the "policy approaches, scientific practices, discourses, and investment strategies that make ecosystems legible, governable, and investable as *systems of critical functions that sustain and secure (certain forms of) human life*" (2022, 87; emphasis in original). Rather than replacing extractivist projects, infrastructural natures "introduce new values that overlap, and are in potential contradiction with, ongoing imperatives to other commodity values."

Infrastructural natures can also be enmeshed with 'gray' infrastructures already in existence, like the spatial fix afforded by Shell Pipeline Company's coastal reclamation projects in south Louisiana. On multiple occasions, Shell has made plans to use oysters for the protection of their oil and gas infrastructure (Kupers 2014). Like SCAPE's Living Breakwaters, Shell's coastal pipeline protection projects also seek to buffer against erosion and stabilize land, but in service of either corporate social responsibility projects (in other words, greenwashing), or as direct protection of increasingly vulnerable fossil fuel infrastructure. The National Petrochemical and Refiners Association would seem to genuflect to the humble oyster in praise of such projects, arguing that oysters will be helping to safeguard national security (Stokes et al. 2012, 8). It lands a little ironically, then, that such projects come in the wake of decades of pipeline and oil platform failures, spilling millions of gallons of oil into oyster reef systems (Stokes et al. 2012, 8).

In August 2021, Shell partnered with Natrx Inc., an adaptive infrastructure company focused on building living shorelines in service of coastal resilience, to introduce barriers along the Louisiana coastline at Hell Hole Bayou. As stated on Shell's website, the twin objectives for this project were "enabling important sustainability benefits, including wetland conservation and oyster creation, while protecting our pipeline infrastructure." The project descriptions from both companies hyperfocus on the language of 'resilience,' a concept that—as so many have argued (Wakefield & Braun 2019; Berlant 2016; Guerin 2019; Scott 2016)—promotes a fantasy of 'sustainable' development while deflecting from continued consolidation of power and accumulation of capital. Working toward more resilient fossil fuel infrastructure certainly does not address the root of, nor neutralize, the problem at hand, but creates the conditions to reproduce it—especially when led by a multinational with a deified self-image that thinks itself capable of the "creation" of oysters, amongst other feats of omnipotence.

When ecosystems become sites for infrastructural investment, they become resources for capitalist production and reproduction. On the one hand, creatures, their processes, and their broader ecological systems appear to be revered and allowed to 'take the lead,' but in these examples it is state and corporate funding that decides when, where, and to whose benefit. This amalgamation of gray and green infrastructures comes to "underwrite 'modern' life" (Wakefield & Braun 2019, 203) as a means of keeping the future at bay (what exactly is being kept at bay depends on whom you ask), playing out as a pattern of attachment that ends up looking a lot like cruel optimism.

Repetitions

Extinction, as a nonlinear process that grips one in a state of anticipation, does funny things to linear time. To narrate histories of extinction is to narrate the end of futures, suggests Dolly Jørgensen (2022, 209). Western scientific comprehension of extinction (as both biological phenomenon and idea) is historically situated and emerged in the context of European colonialism. "From findings of mastodon bones debated by Thomas Jefferson and Georges Cuvier to the rapid disappearance of Steller's sea cows in the northern Pacific, extinction was made visible to Europeans in the New World. The extinction of animals, from the loss of the passenger pigeon and the attempted extermination of the American bison in North America to the intentional eradication of the thylacine of Australia, spread with the expanding white settlement of the nineteenth century" (Jørgensen 2022, 210).

Extinction ostensibly means extinguishment, but the extinct continue to flicker as ghosts enmeshed with land, as reminders (heeded or not) of violent histories, and as haunting threats of extinction anew (Tsing et al. 2017, G6; Hovanec 2023, 250; Tuck and Ree 2013). Our contemporary conjuncture of accelerated extinctions—in which oysters are one casualty—is often marked by hopelessness and despair as suggestions of absolute human spectrality are added to the mix. We sit in the middle of a temporal wobble between intense pulses of carbonic and geopolitical perturbations and the corresponding buffering mechanisms dispersed over long time frames to come, intersecting as a triangulated struggle between individual survival, the 'sustainability' of certain modes of life, and a third, more planetary thing like the desire for mutual flourishing, in which the earth might be more widely regarded as a series of relations. What tends to be occluded in this wobbly intersection, braced against apocalypse, is that this is not the first apocalypse (Whyte 2017).

There is an intimate tether between the decimation of native oyster populations and the extermination of Indigenous peoples (Guerin 2019; McBrien 2016). The imperial habit of, and attachment to, violence is directed toward humans and oysters, and therefore strikes at the relations between them, too, such that Indigenous oyster fisheries which had persisted for millennia prior to colonization were also devastated (Reeder-Myers et al. 2022, 1). To describe this colonial history of destructive repetitions as an attachment is to name it as a mode of being in the world that operates according to the conceit of a universal unfolding (Povinelli 2020), whereby regimes of power are sustained by naturalized habits of violence. The lust for accumulation amasses capital and land, but also extinction. Nonhuman species do not demarcate the edges of that possible extinction; extinction's purview reaches out to the extinguishment of cultures and languages (through direct violence or the insidiousness of assimilation), of peoples (through exhaustion or death), and of the earth (through depletion, extraction, and destruction) (McBrien 2016).

BOP (founded in 2014) published its first land acknowledgment on Thanksgiving, 2023. In a November 23 blog post titled "Acknowledging the Lost History, Untold Stories, and Voices of the Lenape Peoples," BOP shared a statement recognizing that they "are not the original stewards of the land" on which they work, land that is "the territory of many indigenous tribes including but not limited to Canarsie, Munsee Lenape, Lekawe (Rockaway), Schaghticoke, Wappinger, and Matinecock peoples, who were removed from the land by state-sanctioned forced integration, displacement, and genocide." The acknowledgment continues:

> As a group that strives to inspire stewardship through restoration and education, we acknowledge that our work would not exist without colonization. In many ways, the oyster symbolizes the lost history, untold stories, and voices of the rightful stewards of this land, from the past and present. As we work to restore oyster reefs to New York Harbor, we also aim to foster open communication. To ensure that our land acknowledgment is a living document, we look forward to engaging indigenous tribal members in the ongoing drafting process and our work at large.
>
> We commit to center indigenous voices, stories, and learnings in all future community engagement, education, and restoration efforts in New York Harbor. After all, restoration is not just environmental; true restoration happens with people and their practices. (Billion Oyster Project, "Acknowledging the Lost History," 2023)

I wish to reflect on BOP's land acknowledgement because its content, as well as its belated existence nearly a decade after the organisation's founding, archive the specificity of the colonial present in which BOP's work unfolds. On the one hand, their acknowledgement attends to many important truths and names good intentions. BOP's work does indeed take place in a colonial context, responding to a colonial problem. And, as they point out, restoration is not 'environmental' if the definition of environmental rests on a binary distinction between nature and culture. The oyster may well symbolise the lost history, untold stories, and voices of the Lenape people—although the fates of oyster and Lenape are intimately connected, rather than coincidental. But, on the other hand, the land acknowledgement, and its surrounding context, documents the ways that BOP can tend to reproduce the same colonial logics it is ostensibly working to undo. While land acknowledgments may be performative, performativity need not be hollow. In terms of speech act theory, performativity is about enacting or producing something (Butler 1993). Whether that something is *meaningful*, rather than merely gestural, depends on whether the acknowledgment is accompanied by sincere commitment to Indigenous sovereignty and disruption to the structure of settler-colonialism (Lambert et al. 2021). The statement sketches a future for the organization that commits to centering the "voices, stories, and learnings" of Indigenous peoples, but it will be up to BOP to demonstrate that they are working toward the development of institutions and norms that can offer meaningful redress.

BOP's land acknowledgment, while belated, rightfully speaks to the state-sanctioned extermination of the people upon whose land BOP works, and recognizes the need for holistic restoration—inclusive of cultural restoration. But oysters used to remediate water are rendered toxic and inedible, which does not remedy the practices and relations of peoples—on Lenape lands and elsewhere—who cannot presently sustain intergenerational stewardship and food-based relationships. While the social benefits of oyster reef restoration are widely extolled, "these social benefits are still viewed largely through a Western lens" (Gibbs et al. 2023). Structures forged to organize transformation must be built around patterns of existence that can sustain true repair—instead of the endless "management of those who have been made killable, once and future ghosts" (Tuck and Ree 2013, 642).

As of yet, there remains work to be done. On November 28, 2023—so-called Giving Tuesday—a few days after the release of their land acknowledgment, BOP shared an Instagram video that opens with the brief recognition that New York Harbor "was" home to the Lenape people four hundred years ago (implicitly negating a present-day Indigenous claim to those lands as home), before quickly moving on to speak to the loss of the Eastern Oyster. The cause of this loss? "We ate them all," declares Pete Malinowski, founder of BOP. But the "we" who ate them all are not the Lenape or other Indigenous peoples whose relationships of reciprocity allowed oysters to flourish for thousands of years before settlers arrived, colonized, and truly did "eat them all" (Reeder-Myers et al. 2022). Thus, the "we" addressed in BOP's video must presume an audience devoid of Indigenous peoples, flattening its audience into the singular subject of the colonial imaginary and reinforcing Indigenous erasure.

Repair

Oyster death comes to matter in the collective. Individual oyster death may seem to be of little consequence, even to the most stringent of animal ethicists like Peter Singer who, in his 1975 book *Animal Liberation*, "notoriously declared that the line between ethically significant and ethically insignificant animals lies 'somewhere between a shrimp and an oyster'" (Steel 2019, 136). In part, this is because the individual oyster is so radically *other* to the normative grievable subject (Butler 2009). That grievable subject is typically human, and a particular kind of human who fits within the historically contingent and oppressively restrictive category of human (that is to say, one with certain access to wealth and other privileges). We might also extend the definition of 'grievable subject' to the loved ones of that human, including the family dog, or the charismatic polar bear. In contrast, the oyster has a long history of being materially and discursively disparaged, cast as a lowly, dispensable creature against whom the rational human subject might be defined. According to historian Karl Steel, "when a writer wants to argue about the ethical inconsiderability of animals, or to imagine animal life at its most minimal, or to determine the absolute limits of animal life, the animal that offers itself up more readily than any other is the oyster" (2019, 137). This tendency emerged with Plato, writes Steel, and persists to this day.

But what if the radical differences between oysters and humans (the former having neither face nor central nervous system) was a point of recognition that differentiation need not be about separation, but instead connection and commitment (Barad 2010, 266)? What if the losses of oysters, and those further losses patched up or prevented by oysters, illuminated a molluscan ethics of entanglement? Perhaps by way of a sincere grief over oyster loss, the feeling of grief being a recognition of our utterly material interdependencies: the negative space of attachment. In grief, loss becomes disorientation. You stumble, because attachment is shared coordination, and attachments yanked away become a tear in the world.

Like oysters (albeit at a different scale), we humans are fleshy beings tethered to geological substrate. We, too, are flush with salt water that is constantly cycling through our bodies, single pieces of a larger, planetary body of water. If you go sit by the ocean to cry salty tears, you are in the midst of a biogeochemical encounter, affect leaking through you and dripping onto the rocks underfoot. A chemical translation of a feeling into a mineral-rich inundation from your eyes. In the words of the hydrofeminist thinker Astrida Neimanis, such an "aqueous becoming" mediates the unruliness of your fundamentally leaky self, one who is never discrete from the planetary entanglements that shore up your life (2017, 50).

Grief is a process that requires safe ground to move, and move it must. Without the right conditions for unraveling and release, grief (for a loved one, for environments lost, for the worlds we long for) becomes lodged as a knot, with no way to change what is stuck. When oysters embody the possibility of safe ground (cleaner water, sturdier shores, revitalized relation), that possibility can be realized by willing their abundant flourishing. Part of this is something like optimism or hope for ecological salvation squeezed from molluscan tissue. But it is also a practice of entanglement, through different scales of planetary bodies and a kind of shared, planetary nervous system. Oysters and their capacity for attachment have a lot to teach about generosity, adaptation, and what it means to live together. They know how to build homes that others can be safe in, and how to not let go of one another. They know how to collectively devote days and lives to tending the watery worlds within which they are attached. And they know how to hold things steady enough to let things move.

References

Barad, Karen. 2010. Quantum Entanglements and Hauntological Relations of Inheritance: Dis/continuities, SpaceTime Enfoldings, and Justice-to-Come. *Derrida Today* 3 (2): 240–268.

Beck, Michael, et al. 2011. Oyster Reefs at Risk and Recommendations for Conservation, Restoration, and Management. *BioScience* 61 (2): 107–116.

Berlant, Lauren. 2011. *Cruel Optimism*. Durham: Duke University Press.

———. 2012. *Forms of Attachment: Affect at the Limits of the Political*. ICI Berlin. July 9. https://www.ici-berlin.org/events/forms-of-attachment/. Accessed 2 April 2024.

———. 2016. The Commons: Infrastructures for Troubling Times. *Environment and Planning D: Society and Space* 34 (3): 393–419.

Billion Oyster Project. *Ecosystem Engineers*. https://www.billionoysterproject.org/ecosystem-engineers. Accessed 15 November 2023.

———. *Acknowledging the Lost History, Untold Stories, and Voices of the Lenape Peoples*. https://www.billionoysterproject.org/blog/acknowledging-the-lost-history-untold-stories-and-voices-of-the-lenape-peoples. Accessed 24 November 2023.

Burkett, Jeremy R., et al. 2010. Oysters Produce an Organic-Inorganic Adhesive for Intertidal Reef Construction. *Journal of the American Chemical Society* 132 (36): 12531–12533.

Butler, Judith. 1993. *Bodies That Matter: On the Discursive Limits of Sex*. New York: Routledge.
———. 2009. *Frames of War: When Is Life Grievable?* New York: Verso.
Dame, Richard F. 2012. *Ecology of Marine Bivalves: An Ecosystem Approach*. Boca Raton: CRC Press.
Foulon, Valentin, et al. 2018. Proteinaceous Secretion of Bioadhesive Produced During Crawling and Settlement of *Crassostrea gigas* Larvae. *Scientific Reports* 8. https://doi.org/10.1038/s41598-018-33720-4.
Gan, Elaine, and Anna Tsing. 2018. How Things Hold: A Diagram of Coordination in a Satoyama Forest. *Social Analysis* 62 (4): 102–114.
Gibbs, Mitch, et al. 2023. Extending Conservation of Coastal and Oyster Reef Restoration for First Nations Cultural Revitalization. *Conservation Biology* 37 (6). https://doi.org/10.1111/cobi.14158.
Guerin, Ayasha. 2019. Underground and at Sea: Oysters and Black Marine Entanglements in New York's Zone-A. *Shima* 13 (2): 30–55.
Haraway, Donna. 2003. *The Companion Species Manifesto: Dogs, People and Significant Otherness*. Chicago: Prickly Paradigm Press.
Hovanec, Caroline. 2023. Unhoming the Pigeon: Ahmed Ali's Twilight in Delhi. In *Beastly Modernisms: The Figure of the Animal in Modernist Literature and Culture*, ed. Alex Goody and Saskia McCracken, 248–262. Edinburgh: Edinburgh University Press.
Ingold, Tim. 2015. *The Life of Lines*. London and New York: Routledge.
Jørgensen, Dolly. 2022. Extinction and The End of Futures. *History and Theory* 61 (2): 209–218.
Kirby, Michael Xavier. 2004. Fishing Down the Coast: Historical Expansion and Collapse of Oyster Fisheries Along Continental Margins. *PNAS* 101 (35): 13096–13099.
Kupers, Roland. 2014. *Turbulence: A Corporate Perspective on Collaborating for Resilience*. Amsterdam University Press.
Kurlansky, Mark. 2006. *The Big Oyster: History on the Half Shell*. New York: Random House.
Lambert, Michael C., Elisa Sobo, and Valerie L. Lambert. 2021. Rethinking Land Acknowledgments. *Anthropology News*, December 20. https://www.anthropology-news.org/articles/rethinking-land-acknowledgments/. Accessed 2 April 2024.
Lenihan, Hunter S. 1999. Physical-Biological Coupling on Oyster Reefs: How Habitat Structure Influences Individual Performance. *Ecological Monographs* 69 (3): 251–275.
McBrien, Justin. 2016. Accumulating Extinction: Planetary Catastrophism in the Necrocene. In *Anthropocene or Capitalocene: Nature, History, and the Crisis of Capitalism*, ed. Jason W. Moore, 116–137. Oakland: PM Press.

Miossec, Laurence, Rose-Marie Le Deuff, and Philippe Goulletquer. 2009. Alien Species Alert: *Crassostrea gigas* (Pacific oyster). *ICES Cooperative Research Report* 299: 10.17895/ices.pub.5417.

Natrx. *Hell Hole Bayou*. https://natrx.io/case-studies/hell-hole-bayou. Accessed 2 December 2023.

Neimanis, Astrida. 2017. *Bodies of Water: Posthuman Feminist Phenomenology*. London: Bloomsbury Academic.

Nelson, Thurlow C. 1924. The Attachment of Oyster Larvæ. *Biological Bulletin* 46 (3): 143–151.

Nelson, Sara, and Patrick Bigger. 2022. Infrastructural Nature. *Progress in Human Geography* 46 (1): 86–107.

New York State Homes and Community Renewal. *Living Breakwaters: Project Home*. https://hcr.ny.gov/living-breakwaters-project-home. Accessed 27 August 2023.

Povinelli, Elizabeth. 2020. Melting Glaciers, Rising Seas. *Serpentine Galleries: The Shape of a Circle in the Mind of a Fish: The Understory of the Understory*. December 10. https://www.youtube.com/watch?v=OWf8Mv7WATI&ab_channel=SerpentineGalleries. Accessed 2 April 2024.

Reeder-Myers, Leslie, et al. 2022. Indigenous Oyster Fisheries Persisted for Millennia and Should Inform Future Management. *Nature Communications* 13 (2383). https://doi.org/10.1038/s41467-022-29818-z.

Scott, Emily Eliza. 2016. Infrastructure Inside Out. In *Toward an Urban Ecology*, ed. Kate Orff, 186–194. New York: The Monacelli Press.

Shell. Helping to Enhance Biodiversity and Shoreline Stability in Louisiana. https://www.shell.us/sustainability/conservation/conservation-activities/helping-to-enhance-biodiversity-and-shoreline-stability-in-louisiana.html. Accessed 20 November 2023.

Steel, Karl. 2019. *How Not to Make a Human: Pets, Feral Children, Worms, Sky Burial, Oysters*. Minneapolis: University of Minnesota Press.

Stokes, Shawn, et al. 2012. *Restoring Gulf Oyster Reefs. Center on Globalization, Governance & Competitiveness*. Duke University. http://www.oyster-restoration.org/wp-content/uploads/2012/06/CGGC_Oyster-Reef-Restoration.pdf. Accessed 2 April 2024.

Tsing, Anna, et al., eds. 2017. *Arts of Living on a Damaged Planet: Ghosts and Monsters of the Anthropocene*. Minneapolis: University of Minnesota Press.

Tuck, Eve, and C. Ree. 2013. A Glossary of Haunting. In *Handbook of Autoethnography*, ed. Stacey Holman Jones, Tony E. Adams, and Carolyn Ellis, 639–658. Walnut Creek: Left Coast Press.

Wakefield, Stephanie, and Bruce Braun. 2019. Oystertecture: Infrastructure, Profanation and the Sacred Figure of the Human. In *Infrastructure, Environment, and Life in the Anthropocene*, ed. Kregg Hetherington, 193–215. Durham: Duke University Press.

Wells, H.G. 1891. Zoological Retrogression. *The Gentleman's Magazine* 271: 246–253.
Whyte, Kyle. 2017. Our Ancestors' Dystopia Now: Indigenous Conservation in the Anthropocene. In *The Routledge Companion to the Environmental Humanities*, ed. Ursula Heise, Jon Christensen, and Michelle Niemann, 206–215. London and New York: Routledge.
Wilbon, Anthony, Mark Bundy, and Kelton Clark. 2011. Entrepreneurship in the Chesapeake Bay Oyster Industry. *Journal of Case Studies in Management* 9 (2): 1–15.
Wright, Kate. 2014. Becoming-With. *Environmental Humanities* 5 (1): 277–281.

CHAPTER 7

Surreal Seas and Embodied Encounters: Elizabeth Bishop's Darwinian Poetics

Karen Eckersley

At the beginning of Elizabeth Bishop's prose poem "Strayed Crab," from her collection *Rainy Season; Sub-Tropics* (1969), the crustaceous speaker observes that they are not at home asking, "How did I get so far from water?" (Bishop 2011, 163). This chapter examines how Bishop bridges the gap noted by her crab speaker, taking the reader back to the water both through her interconnective meditations on the lives of three nonhuman seaside dwellers and, most crucially, via her sense of our Darwinian genealogical connection to them. I explore the ways in which the ocean is a conspicuous presence in Bishop's work where her sea-infused writing brings readers closer to the water through a series of nonhuman encounters.

Human aquatic ancestry is intimated through the watery worlds of the marine creatures evoked in the poem, enabling the reader to confront both the oceanic animal they once were as well as to navigate the shoreline from the perspective of a crab, snail, and toad. The processual and metamorphic ontologies that Bishop bodies forth in her writing with Darwinian precision reorientate humankind's lens to recognise the vital essence of

K. Eckersley (✉)
Nottingham Trent University, Nottingham, UK
e-mail: karen.eckersley2019@my.ntu.ac.uk

© The Author(s), under exclusive license to Springer Nature Switzerland AG 2024

V. Fibisan, R. Murray (eds.), *Blue Extinction in Literature, Art, and Culture*, Palgrave Studies in Animals and Literature,
https://doi.org/10.1007/978-3-031-69910-8_7

our own watery inheritance. I posit that by bringing us back to the marine realm that we have become alienated from, Bishop's sea-borne speakers in *Rainy Season; Sub-Tropics* have the capacity to help humankind confront an anthropogenically driven climate crisis that continues to privilege land-based narratives. More specifically, this chapter will consider the vision of embodied encounters fostered throughout the three poems, positing that they intimate a recuperative strategy for reimagining our relationship with the benthic world at a time of environmental collapse. I argue that Bishop's prose-poems present three forms of embodied encounter: firstly, with our own blue genealogical heritage; secondly, between human and marine nonhuman species; and finally, through a spatial encounter that bodies forth the necessarily co-dependent dynamic between land and sea. In these interconnective contact zones I argue that Bishop fosters a vision where we are no longer "so far from water," bringing human and nonhuman seascapes into conversation in a way that underlines the entangled, oceanic environment that implicates us all.

Bishop wrote *Rainy Season; Sub-Tropics* in January 1967 when she was briefly reunited with her lover Lota de Macedos Soares in Brazil. Their relationship had broken down, but the poems were written during a lull in their breakup. First published in *The Kenyon Review* in November 1967, the three poems document the experiences of a Giant Toad, Strayed Crab, and a Giant Snail. There has been relatively little scholarship conducted upon these poems despite the burgeoning field of animal studies, and the presence of nonhuman—and specifically blue species in these poems—warrants further investigation. Zhou Xiaojing writes of Bishop's dialogic investigations on subjectivity in the poem (Xiaojing 1994, 75–92), while Lorrie Goldensohn offers an insightful reading of the collection, exploring its "psychic dilemmas" in what she sums up as a "more pointed response to the treacheries of self-representation" (Goldensohn 2014, 191). In her book *Elizabeth Bishop's Brazil* (2016), Bethany Hicok discusses *Rainy Season* in the briefest of terms in a footnote arguing that the "oversized and misplaced creatures are obvious alter egos for the poet and her art" (150). While I broadly agree with Hicok in recognising the migratory habits of these animals as well as their discomfort, I do not share her thoughts on the fact that these waterside creatures are mere representatives for Bishop herself. Rather, this chapter argues that Bishop's immersive seascapes, seen emphatically in *Rainy Season*, can be read as anticipatory examples of the ways in which we can connect and engage with the ocean-scape rather than remain alienated from it in this time of impending environmental catastrophe.

Stacy Alaimo's examination of our own watery ancestry as a means of reconnecting with an environment that is populated with our own kin, rather than a presumed alien other, advances ecological themes that resonate with Bishop's own work:

> Submersing ourselves, descending rather than transcending, is essential lest our tendencies toward Human exceptionalism prevent us from recognizing that, like our hermaphroditic, aquatic evolutionary ancestor, we dwell within and as part of a dynamic, intra-active, emergent, material world that demands new forms of ethical thought and practice. (Alaimo 2011, 283)

The submersive mode that Alaimo calls for in order to confront and potentially challenge humankind's supposed transcendence both travels in the direction of our own prehistoric past as well as "towards" our ongoing kinship with aquatic species. Such an immersive approach, seen in Bishop's seascapes and most redolently in *Rainy Season*, has the potential to overcome repeated visions of the ocean as a place the human species feels disconnected and thus alienated from. It is important to add that Bishop's own sense of kinship with the sea is rooted in biographic detail: Bishop grew up mainly in Nova Scotia and spent much of her adult life near the sea. Being close to the water, whether in Key West, Rio de Janeiro, or Boston, meant that the ocean played a central role in her poetry. The word "sea" features prominently in her poems, speaking to the fact that the seascape was a key detail of her psychic and poetic geography.

In a recent essay, Sarah Kennedy has identified what she describes as "transitivity and intra-species empathy in the poems' phenomenological framing" (Kennedy 2022, 118). Kennedy's reading speaks to both Alaimo's approach and to the reading that I am exploring here. As Kennedy notes in her study of the collection, all three nonhuman animals, at least initially, appear to be out of place in some way, despite the fact that they reside in their sub-tropical homeland. In the first poem the Giant Toad explains, for example, that he is "too big, too big by far" (Bishop 2011, 163) while the Strayed Crab, as mentioned earlier, asserts that he is in a place that is "not my home." And yet these spaces are indeed in their natural habitats where Bishop invokes primordial geologies of mist-laden proportions and lichen-covered rocks that engender the geographies within which they reside. Here, Bishop's poem causes us to encounter the ecology of our own beginnings, traversing the strange and yet familiar marine milieu that recalls Darwin's observations about human oceanic inheritance and genealogy:

> Our ancestor was an animal which breathed water, had a swim bladder, a great swimming tail, an imperfect skull, and undoubtedly was a hermaphrodite. Here is a pleasant genealogy for mankind. (Neimanis 2017, 109)

Darwin's cheerful tone notes the human and sea-creature interconnection that characterises our own beginnings as a species, anticipating Alaimo's argument and thus Bishop's poems. Such a "pleasant genealogy" helps to situate Bishop's prose-poem as one of encounter both with the creatures she invokes but also in the way that their lineage can be traced back to one where we emerged as a species. Timothy Morton urges in a way that Bishop's poems communicate: "Just read Darwin. Evolution means that life-forms are made of other life-forms" (Morton 2010, 275). While the animals seen in this prose poem do at first seem to evoke a sense of displacement and indeed have an alien quality to us in their exoskeletal and amphibious physicality, they at the same time exhibit a surreal familiarity. The speaker's tone naturalises this link between humans and crabs rather than presenting it as something unnatural. Their physiology is no more bizarre to our human eyes than the spectacle of our own "swimming tail" and "swim bladder" of the past that Darwin reminds us of, but all the while Bishop etches these characters as living, breathing ontologies who navigate the same fears and concerns that we may have done and indeed still do. While elements of their physiology are characterised in a way that may resonate with their own kind, I also posit that their occasional sense of estrangement from the watery space they occupy may also mirror the way in which we similarly experience a disconnect from the seascape in this current period of climate crisis. Through an affirmative tone Bishop fosters instead a poetics of immersion, connection, and encounter where we reconnect across species, time, space, and sea. This is especially clear in the final poem featuring the Giant Snail whose difficulties seem to ease once he is back in the water that the opening speaker had been searching for:

> If and when I reach the rock, I shall go into a certain crack there for the night. The waterfall below will vibrate through my shell and body all night long. In that steady pulsing I can rest. All night I shall be like a sleeping ear. (Bishop 2011, 165)

There is a return to origin in this bodily connection that the snail experiences here; we see a physical encounter between creature and land but also to his beginnings that he finds in the folds of the land and sea edge. I argue that such an encounter of sensory depths speaks to the watery ancestry that Bishop articulates in her poetry where to be by the sea and enmeshed within it is to find a deeply sensorial interconnection that speaks to the world from whence we came as a species. There is a palpable synaesthesia at work here, in which the snail's hyper-responsiveness to his environment melds sound, vision, and touch. There is a melting of boundaries, a conflation of creature and environment, as well as of our perceptions of them as readers, with Bishop exerting an inter-connective dynamic, exploding the imagined boundaries between land and sea.

Kirstin Hotelling Zona articulates these creatures' positions succinctly when she argues that Bishop's poetry is preoccupied with "probing the complex and often discomforting nature of self as it is shaped through interaction with the world" (Zona 2014, 49). This can be seen in the compelling synaesthesia communicated via Bishop's description of the snail, as if the sensual entanglement he finds within this seascape fosters a closeness that induces the comforting sensations of his own origins and thus the phenomenological relationship with this environment. It is a dynamic that Bishop fosters in her poetry as significant to all species and aquatic species, especially given that we tend to feel so distant from them. To speak of the snail and his counterparts by the ocean, then, is also to speak of our own oceanic heritage as articulated by Darwin, where the sea is no longer an alien realm from which we are estranged but rather a space of kinship. The snail's sleeping ear and "steady pulsing" as he resides above the waterfall within this seascape invokes the "pleasant genealogy for mankind" (Neimanis 2017, 109) that Darwin celebrates in his studies. The vibrations, pulsing, and the soundscape that Bishop evokes is suggestive of a gynocentric realm—a womb-like return that the folds of the sea seem to suggest in her prose-poem.

The navigational challenges and experiences of the three creatures are conspicuously difficult at times but there is nevertheless a deeply entrenched sense of belonging and attachment to which Bishop's speakers return. This sense of belonging positions us as readers, unlike the strayed crab, much closer to the water than we might anticipate. What Bishop demands of her readers is for us to fully reconnect with this seascape and its inhabitants, as Kennedy argues here:

> the poetic consciousness engages in repeated acts of imaginative self-transposal, speaking—insofar as is possible—from within an animal-shaped experience as describing a world mediated by (imagined) animal sensory experiences and projections [...] Their concern [is] with feeling out—as best they might—the contours of toad, crab and snail from the inside. This provides a rationale for Bishop's choice of animals: the toad's amphibious skin and the exoskeletons of crab and snail provide distinctively non-human models of phenomenological experience. (Kennedy 2022, 124)

Kennedy's analysis alludes, in part, to the fact that Bishop's three poems are populated by speaking animals that are "feeling" their way forward. These represent moments of what she describes as "imaginative self-transposal" where the reader is able to forge an empathetic connection with a seaside creature who might at first glance appear consciously alien in its physiological make-up. We must connect with these animals in a cerebral sense but also in a physical way, where their "animal sensory experiences," as Kennedy puts it, communicate the world to us in a phenomenological and emotional manner. In the opening poem this is most conspicuous when the giant toad character is subjected to the cruelty of the human in a way that may well engender the reader's sympathy as they recognise the bullying traits of our species. Indeed, the toad implores the reader: "Pity me,"

> I live, I breathe, by swallowing. Once, some naughty children picked me up, me and my two brothers. They set us down again somewhere and in our mouths they put lit cigarettes. We could not help but smoke them ... I thought it was the death of me ... I was sick for days. (Bishop 2011, 163)

The toad identifies himself and his place in the world in this example, breathing as we do and yet prey to the cruelty of the human species as this episode demonstrates. In one sense Bishop presents this most alien of creatures as like us in his respiratory needs and the fear he experiences during these moments of bullying. As a reader we are pulled towards this speaking animal, connecting with his suffering despite the bodily strangeness of an amphibious creature who is orientated towards the sea. Bishop's poem suggests that the toad not only feels significant pain, but worse, he hurts at the hands of human species. The poem represents an intense encounter both between toad and human children in the poem itself but also between the toad and readers, who find themselves entangled in the plight of the alien creature via the empathy it stirs.

Bethany Hicok explains that the toad is one of Bishop's "oversized and misplaced creatures" who function as "alter egos for the poet and her art" (Hicok 2016, 150). This observation speaks to the sense of intense connection that Bishop's poem appears to be communicating. However, the speaking animal episodes we witness are also an exercise in the surreal familiarity that aligns Bishop's poetic practice with Darwin's own observational studies. Bishop's poetry borrows Darwin's cheerful tone of observation, tracing similarly convivial lines between humankind and nonhumans as well as our aquatic ancestry. I suggest that such modes of surrealism prompt a recalibration of our own relationship with nonhuman others and the marine milieu in particular. As Kennedy explains, Bishop's work is "full of oblique angles: the tilt, the skew, the incline" (Kennedy 2022, 126). My sense is that Bishop captures the unsettling visions of the world in a way that she observes in Darwin's own practice. Bishop articulates her admiration of Darwin's approach in her letter to Ann Stevenson, remarking of his writing's

> glimpses of the always-more-successful surrealism of everyday life, the unexpected moments of empathy (is it?), catch a peripheral vision of whatever it is one can never really see full-face but seems enormously important. (Bishop 2008, 861)

I argue that the importance of our connection with the nonhuman world and indeed our nonhuman ancestry—particularly at sea—is what Bishop identifies in her poems as "enormously important." In her prose poem she attempts to examine this vision of our human and nonhuman ontology "full-face" in a way that is unsettling, strange, and yet necessary in order to recalibrate our place as a species within the world rather than atop of it. Like Darwin's recorded observations we view the natural world we are immersed in as "aslant," "unexpected," and as Susan McCabe argues "filled with attractive anomalies and 'originals'" (McCabe 2009, 548).

As I have posited, the animals in Bishop's aquatic prose poems articulate an acute self-awareness but also a conspicuous identification of their position in relation to others. They are bodies that pitch themselves in the context of the coastal environment and the creatures around them, speaking to an amalgam of encounters that the human eye can be blind to in its hierarchical approach to the world. The Giant Toad asserts: "Don't breathe until the snail gets by" (Bishop 2011, 163). However, he acknowledges that "we go travelling the same weathers" (Bishop 2011, 163).

Here, we witness a vigilance towards another species while at the same time a recognition that their experiences resonate and are to a certain extent aligned. The toad also sends out an alert to the crab warning him to watch out for his own poison: "If I will it, the poison could break through, blue-black, and dangerous to all. Beware, you frivolous crab" (Bishop 2011, 163). These lines, interestingly, precede the crab's own arrival in the second prose poem as if his existence is seen immediately in the context of another. In the second prose poem the crab identifies himself via his strange physicality, where Bishop charts his aesthetics in Darwinian detail:

> I am the color of wine, of *tinta*. The inside of my powerful right claw is saffron-yellow. See, I see it now; I wave it like a flag. I am dapper and elegant; I move with great precision, cleverly managing all my smaller yellow claws. I believe in the oblique, the indirect approach, and I keep my feelings to myself. (Bishop 2011, 163-4)

The crab exhibits the acute self-awareness that we witness in the giant snail. He documents himself with the precision of a natural historian inviting us to really "[s]ee" him, perhaps as we have never seen this creature of the sea before. His belief in the "oblique" and his movement across the environment sideways is suggestive of the way that Bishop presents things aslant, or differently in order to articulate other, even nonhuman ways of seeing. He is "indirect" and yet he is also forthcoming in the way he experiences himself, the world, and the other sea creatures around him. In the same way that the crab tells him to "beware," he is also mindful of the snail perceiving him as a "big soft monster, like a yellow cloud" (Bishop 2011, 164). Indeed, the crab navigates the seascape with the snail very much on his radar, as the following demonstrates:

> What is it doing? It pats my pack. Out, claw. There I have frightened it away. It's sitting down, pretending nothing's happened. I'll skirt it. It's still pretending not to see me. Out of my way, O monster. (Bishop 2011, 164)

While the crab exhibits fear posturing as aggression, he is deeply curious of the snail experiencing his presence as a pat on the back early on and pondering, very clearly, what he is doing. Ultimately the crab wants the snail to "Cheer up," tapping his shell "encouragingly" (Bishop 2011, 164). In the next section the crab again observes the parallels between himself and the toad who is similarly cast as a "watchdog" (Bishop 2011, 164). The giant snail demonstrates how he is in conversation with the crab, pondering inquisitively: "What's that tapping on my shell?" (Bishop 2011, 165). The snail exhibits a deep kinship with his toad neighbour, suggesting that there is a physical, and indeed emotional synchrony between these two creatures of the sea: "That toad was too big, too, like me. His eyes beseeched my love. Our proportions horrify our neighbours" (Bishop 2011, 165). Here we witness not only a sensitive awareness of another species, but also the capacity of the two to sympathise with another and so find connections. Kennedy explains that the poems are in fact propelled principally by this interrelationship:

> All of the three poems are obsessively preoccupied with the dynamics of self and other. They purport to record the perceptional experience of subjective awareness of a non-human subject. Each poem combines minute self-examination with an externalising self-consciousness that projects itself onto an imagined community of observers. (Kennedy 2022, 122)

All three creatures are pitched as living by their senses, sharpened as they are in relation to each other and as a result of the ecologies within which they are enmeshed. They seek and find connection, even in circumstances when their ontologies would seem to estrange them. We ultimately witness these creatures as sharing a sense of boundlessness, in part in the way that they encounter each other as fellow beings of the seaside, but also in the way that they meld with the watery environment that holds them. In this way they anticipate the boundless materialisms that Karen Barad celebrates, exhibiting a sort of "exteriority within" (Barad 2008, 122). Indeed, it is through interaction with others and places, or "intra-actions" as Barad articulates it, that the Cartesian privilege of the human is brought to ground. As she explains it, all "relata do not pre-exist relations; rather, relata-within-phenomena emerge through specific intra-actions" (Barad 2008, 133). In the case of Bishop's *Rainy Season* creatures, the visions of encounter we witness in the poems are examples of such vitalist "intra-actions" that implicate the human and nonhuman in a way that is non-hierarchical. As Barad argues: "'Human' bodies are not inherently different from 'nonhuman' ones" (Barad 2008, 141). In a way that we see played out in Bishop's prose poems, Barad insists that "[b]oundaries do not sit still" (Barad 2008, 135) and so "[a]ll bodies, not merely 'human' bodies, come to matter through the world's iterative intra-activity" or through its encounters. It is these embodied encounters that bring the oceanic realm to the fore in this prose poem shifting the land-based bias that has fostered a more privileged green ecological agenda towards recognising the equally vital significance of the blue milieu from where we derive, and which we depend upon.

The boundlessness that Barad advances in recognising a non-hierarchical and so posthuman approach to the world is abundant in Bishop's sea-borne poetry and most specifically in *Rainy Season; Sub-Tropics.* Bishop immerses readers in a poetic and virtual seascape of slippages and movement, where the creatures slide between invocations of land and water that meld organically together. With the strayed crab we enter a "pool," lodged as it is between land and water, inhabited by "skittering waterbugs" (Bishop 2011, 164) who navigate the water's surface with delicate accuracy. They are like the headline creatures of the prose poems themselves who are all equally adapted to land and water, navigating the shore with tentative care. As the crab explains, Bishop's poetry admires "lightness, and agility" (Bishop 2011, 164) seeing them as rare qualities in this "loose world." This fluid vision of the nature of existence is similarly captured in Bishop's evocation of the waterfall, tumbling as it does from the land that is comprised of "the smallest stones and sticks," as the snail notes, as well as "mountains of black rock" (Bishop 2011, 164). It is this eschewal of fixity that characterises all three creatures in their sea-borne adventures, where slipping between space and time speaks to the malleable nature of their existence, and their environment. Their world echoes the one that Barad notes in her observation that "[b]oundaries do not sit still."

Bishop's choice of the prose poem form further underscores the themes of hybridity and fluidity that she communicates through the text. The prose poem navigates between two worlds, presenting a slippage of forms that eschews epistemological fixity in a way that Barad's posthuman politics would advance. Clive Scott argues that the form engenders a freedom that I would suggest enables the boundlessness we witness in Bishop's collection:

> [The prose poem is] a site of poetic transformation, where one kind of verse structure is converted into another, or where prose acts as an embedding medium in which prosodic structures are free to imagine and constantly reimagine themselves. (Scott 1990, 119)

Scott's analysis aptly demonstrates how Bishop's poem has the capacity to catalyse a blue ecological reimagining: it enables a malleable approach that shakes off humanist hierarchies that privilege boundaries, collapsing divisions, and embracing immersion and encounter. The prose poem's capacity to function as an "embedding medium" is an important feature that we witness in Bishop's poem, where she fosters an approach that also embeds humankind within the seascape she creates, melding together land and sea, species and environment, ancient and current temporalities. It is this openness to the world that Bishop fosters, one where the creatures voice an awareness of others and their environments as well as a nomadic sensibility that eschews the fixity of humanism that has only estranged us further from an oceanic realm that our survival depends upon. The prose poem facilitates the mode of encounter that most aptly characterises Bishop's writing. Within these hybridised forms Bishop enables nonhuman creatures to enter into a form of virtual conversation and encounter; while each prose poem appears to be dedicated to an individual animal, others are able to appear and interact with them. This crossing of boundaries, both of poetic form and geographically in the way that the creatures exert a slippage between land and water, invites communion with others and inclusivity. It invites the human and nonhuman to coexist together in seascapes that implicate all ontologies in a collective dynamic necessary for the cause of climatic recuperation and recovery. In this way the prose poem is a tool that has the potential to help us navigate back to the waterside from which we have become so estranged.

Bishop's prose poem is interested in encounters not only across space but also across temporalities. While the environmental concerns we are currently grappling with are anticipated in the poem's themes, I would also argue that the poem explores the primordial ancestries and seascapes with which we hold a deep bodily connection. Bishop's writing situates the reader, and her creatures, within a conspicuously prehistoric sphere populated by ancient geologies. For example, the "Giant Toad" character imagines his own "mottled hide" as "silver on a mouldering leaf," while the lichen itself, a frequent feature of Bishop's poetry, is described as "gray," perhaps asserting its longevity (Bishop 2011, 163). There is a spectral quality to the poems, haunted as they are by the ancient ecologies that pervade the creatures' environs, and speaking to the way that this watery environment endures and implicates us in our current climatic predicament. Bishop, I suggest, is reminding us that this seemingly ancient seascape still has something to say and has resonance in our modern world:

> The mountains of black rock give off such clouds of steam! Shiny streamers are hanging down their sides. When this occurs, we have a saying that the

> Snail Gods have come down in haste. *I* could never descend such steep escarpments, much less dream of climbing them. (Bishop 2011, 164-5)

The "clouds of steam!" evoke a primordial past, while the snail's acknowledgement of his species' "Gods" is a similar nod towards his ancient beginnings. It is a demonstrable awareness of the history of which he is comprised, a vital component that is a part of his bodily makeup over time. Further references to his own evolution later in the poem are reflected by his "four horns that can't attack" and the fact that the "sides of my mouth are now my hands" (Bishop 2011, 165). These adaptations over time, an allusion to the microscopic observations recorded by Darwin, speak to our own relationship with the past as well as our own bodily connection in that it similarly orientates the human species towards the sea. The snail's visible adaptations and shifts over time are articulated as a part of his make-up rather than a discrete and estranged incarnation of him that is compartmentalised in the past. To encounter the snail is to encounter the evolved and adapted creature he has become as well as the past from which he has travelled. I argue that Bishop's recurrent acknowledgement of ancient geologies and primordial spheres, as well as the creatures' physical adaptations, similarly demand that we too as a species recognise our own watery past. In these prose poems we encounter this version of ourselves through the crab, snail, and toad. Such recognition is consolidated further through the way in which Bishop bodies forth these creatures as talking animals. This is clearly an example of anthropomorphising, but rather than reducing these creatures to tokenistic vehicles of humanist articulation, Bishop's anthropomorphism has the capacity to bridge the gap between species, and particularly oceanic species. Kari Weil discusses the topic of anthropomorphism in a way that speaks to the potentiality of Bishop's speaking sea creatures in the poem:

> On the one hand, as a process of identification, the urge to anthropomorphize the experience of another, like the urge to empathize with that experience, risks becoming a form of narcissistic projection[.] ... On the other hand, as a feat of attention to another and of imagination regarding the other's perspective, this urge is what brings many of us to act on behalf of the perceived needs and desires of an other/animal. (Weil 2012, 19)

What Weil describes here in the second part of her quotation communicates the practices of Bishop in *Rainy Season*, where her poetic and bodily invocations of the creatures forge a powerful and imaginative empathy that bridges the gap between species, space, and arguably temporality itself. This is not the self-indulgent articulation of a poet who purportedly eschewed the practices of confessional poets. Indeed, Bishop was explicit about the fact she disliked confessional poetry as a form, explaining in *Confessional Poetry* (1967) that some of the poets who employed these practices had a tendency to "overdo the morbidity," and that she wished "they'd keep some of these things to themselves" (Goldensohn 2014, 191). Rather, as Lorrie Goldensohn underlines, Bishop aims to satirise self-importance, in turn emulating Darwin's direct and observational voice on the nature of things and the things of nature. Bishop famously wrote of her admiration for the young scientist "sinking or sliding giddily off into the unknown," and she seeks to achieve a similar result in her own poetry, explaining: "What one seems to want in art, in experiencing it, is the same thing necessary for its creation, a self-forgetful, perfectly useless concentration" (Bishop 2008, 861). These remarks speak both to Bishop's admiration of Darwin and her wish to eschew the indulgent and self-conscious thought process in order to lose herself to the exterior realities of the environment in which she is writing and observing.

Bishop emphasises this observational praxis in her poetics further in a letter to Jerome Mazzuro, an author and literary critic. Responding to his observations that her poem "The End of March" contains literary allusions, she argues that most importantly her writing is grounded in observation:

> It takes an infinite number of things coming together, forgotten, or almost forgotten, books, last night's dream, experiences past and present—to make a poem. The settings, or descriptions, of my poems are almost invariably just plain facts—or as close to the facts as I can write them. (Bishop 1994, 621)

With these remarks in mind, I want to suggest that Bishop's recourse to an anthropomorphic method for the creatures is less an example of her ventriloquising her own desires or needs, and more an example of the ways in which she observes the nonhuman as a species among others that suffer, desire, and overcome in the same manner that we do as a human species. Bishop's anthropomorphism is an act of empathetic interconnection across boundaries that exemplifies the vital importance of embodied encounter in her poetry as a means of exploding the boundaries that humanism insists upon. In this way Bishop's anthropomorphic speaking animals act as a sort of navigational tool rather than as a reassertion of humanism's presumed hierarchical self-importance. Indeed, it is not about Bishop at all or the human at the top of the species but rather an articulation of the way that we are one among others, and most importantly others situated within the aquatic as much as the terrestrial environment. *Rainy Season* thus anticipates "a notion of the post-human that acknowledges human animals as having coevolved with innumerable species without whom we would not be who we are and with whom we share our environment and resources" (Weil 2012, 18).

Sidney Dobrin articulates our entanglement with the ocean with a clarity and directness that underpins Bishop's oceanic poems: "Ocean is not an object, a body, a place, a concept or a way of being I try to get to but is something from which I can never detach" (Dobrin 2021, x). As he further explains, the "'bodies' of saltwater that cover the planet are connected … a single aquatic body divided only by human cartography and discourse" (Dobrin 2021, 1). This chapter has underlined the ways in which Bishop's prose poetry reunites us species of the sea through her representation of interactive encounters where the toad, snail, and crab intermingle as species. I have also underscored the ways in which these seemingly alien animals meld with their environments in a way that we once did as aquatic ancestors of the seascape. This chapter has demonstrated how the spectre of our sea-borne past emerges in these prose poems, speaking to both a re-immersion within the oceanic realm as well as its invocation of boundlessness that explodes humanist boundaries. By immersing ourselves in this poetic invocation of the sea with these creatures, Bishop stages a vital return to the ocean where we are no longer so far from water in the way that concerns the toad in the opening poem. Through the intimacy of these bodily, ontological, and environmental encounters Bishop re-immerses us as a species within the seascape, rekindling our own entanglement within it as we confront a climate crisis that is playing out in the ocean as much as on land.

References

Alaimo, Stacy. 2011. New Materialisms, Old Humanities, or Following the Submersible. *NORA – Nordic Journal of Feminist and Gender Research* 19 (4): 280–284.

Barad, Karen. 2008. Posthumanist Performativity: Toward and Understanding of How Matter Comes to Matter. In *Material Feminisms*, ed. Stacy Alaimo and Susan Hekman, 120–154. Bloomington: Indiana University Press.

Bishop, Elizabeth. 1994. *One Art: Elizabeth Bishop Letters*. Ed. Robert Giroux. London: Chatto & Windus.

———. 2008. *Poems, Prose, and Letters*. Ed. Robert Giroux and Lloyd Schwartz. New York: Library of America.

———. 2011. *The Centenary Edition Poems: Elizabeth Bishop*. London: Chatto & Windus.

Dobrin, Sidney I. 2021. *Blue Ecocriticism and the Oceanic Imperative*. London and New York: Routledge.

Goldensohn, Lorrie. 2014. Bishop's Posthumous Publications. In *The Cambridge Companion to Elizabeth Bishop*, ed. Angus Cleghorn and Jonathan Ellis, 183–196. Cambridge: Cambridge University Press.

Hicok, Bethany. 2016. *Elizabeth Bishop's Brazil*. Charlottesville and London: University of Virginia Press.

Kennedy, Sarah. 2022. "Swerving as I Swerve": Elizabeth Bishop's Fugitive Empathy. In *Reading Elizabeth Bishop: An Edinburgh Companion*, ed. Jonathan Ellis, 117–131. Edinburgh: Edinburgh University Press.

McCabe, Susan. 2009. Survival of the Queerly Fit: Darwin, Marianne Moore, and Elizabeth Bishop. *Twentieth Century Literature* 55 (4): 547–571.

Morton, Timothy. 2010. Guest Column: Queer Ecology. *PMLA* 125 (2): 273–288.

Neimanis, Astrida. 2017. *Bodies of Water: Posthuman Feminist Phenomenology*. London: Bloomsbury Publishing Plc.

Scott, Clive. 1990. *Vers Libre: The Emergence of Free Verse in France, 1884–1914*. Oxford: Clarendon Press.

Xiaojing, Zhou. 1994. "The Oblique, the Indirect Approach": Elizabeth Bishop's *Rainy Season; Sub-Tropics*. *Chicago Review*. 40 (4): 75–92.

Zona, Kirstin Hotelling. 2014. Bishop: Race Class, and Gender. In *The Cambridge Companion to Elizabeth Bishop*, ed. Angus Cleghorn and Jonathan Ellis, 49–61. Cambridge: Cambridge University Press.

Weil, Kari. 2012. *Why Animal Studies Now?* New York and Chichester, West Sussex: Columbia University Press.

CHAPTER 8

Hydromaterialism and Membrane Logic in Elizabeth-Jane Burnett's *of Sea*

Vera Fibisan

Invertebrate Encounters

Elizabeth-Jane Burnett is a UK-based ecopoet and academic of Kenyan and British heritage who has published two poetry collections, *Swims* (2017) and *Of Sea* (2021), as well as two works of creative non-fiction, *The Grassling: A Geological Memoir* (2019) and *Twelve Words for Moss* (2023). Burnett's most recent collection, *Of Sea*, builds on her previous published work and methods of environmental exploration. The focus lies on encounters with specific invertebrates in a local body of water. Burnett's collection is also a response to the Sixth Extinction, a time of accelerated anthropogenic biodiversity loss, in which the health of the world's water is under threat—and consequently, the health of every being on the planet. Most affected are invertebrate species, which are often overlooked in favour of charismatic megafauna, and Burnett's collection's focus on aquatic invertebrates renders it a ground-breaking text that pushes the limits of lyric form to rebel against readily received aesthetics. The experimental free verse text was generated out of a practice of swimming, or swimwork, a

V. Fibisan (✉)
University of Sheffield, Sheffield, UK

© The Author(s), under exclusive license to Springer Nature Switzerland AG 2024

V. Fibisan, R. Murray (eds.), *Blue Extinction in Literature, Art, and Culture*, Palgrave Studies in Animals and Literature,
https://doi.org/10.1007/978-3-031-69910-8_8

method that arises in the work of other contemporary writers, such as Roger Deakin's *Waterlog* (1999), and Victoria Whitworth's *Swimming with Seals* (2017).[1] Burnett's *Of Sea* shows us that literature can bridge the gap between a polluted world and the human self, and that the lyric form can contribute towards facilitating connections to the more-than-human world in crisis and opening multispecies dialogues through its employment.

Hydromaterialism brings together Stacy Alaimo's concept of "new materialism" and Astrida Neimanis's notion of "hydrofeminism."[2] These two core components of hydromaterialism form an eclectic amalgam that refines avenues of thought with regards to human behaviours in a world in crisis. I will posit these concepts against aquatic biodiversity loss or blue extinction, as seen through Burnett's collection *Of Sea*, with particular emphasis on the body as an assemblage of organised *matter*. Neimanis draws on Nancy Tuana's notion of "viscous porosity" (Neimanis 2016, 91) to define membrane logic as the ability of water to traverse bodies that are nevertheless held together by permeable membranes. Hydromaterialism acknowledges the movement, travel, and limitations of water within two different yet related instances of membrane logic: that of the body and its permeable boundaries enacted by the barrier of skin, and that of the hydrocommons in which water circulates through the hydrosphere. Burnett's lyric encompasses both instances of membrane logic. Within the context of blue extinction explored through swimwork, membrane logic operates as the binding agent between polluted bodies and environments.

This chapter will examine the synergies between what I term hydromaterialism, and membrane logic via three investigative avenues. Firstly, I will develop the theory of hydromaterialism and its position within an existing body of research, drawing on Burnett's *Of Sea* to highlight the ways in which water can facilitate extinction discourses. I will then explore Burnett's practice-based approach of swimwork and the benefits that it affords to portraying liminal spaces in lyric form. Additionally, I will focus on the risks posed by polluted waters to the writer as well as the invertebrates living there, and the dynamics of this shared space. To conclude, I will look at scale and deep time, placing Burnett's collection of detailed encounters against a global backdrop of aquatic extinction and a polluted hydrocommons.

[1] I utilised this term in my doctoral thesis (2020) to describe the type of creative practice of writers such as Elizabeth-Jane Burnett.

[2] See Alaimo (2008, 2011, 2012, 2014); see Neimanis (2012, 2022).

A *Matter* of Life and Death

Of Sea gathers invertebrate "encounters through swimming" (x), presenting them from a non-hierarchical perspective which sees the human body immersed and lowered in a body of water, "descending rather than transcending" (Alaimo 2017, 112).[3] These encounters include intertidal marine creatures as well as terrestrial ones living near water such as bees, beetles, butterflies, moths, flies, aphids, and other insects. I will focus primarily on the marine invertebrates encountered in this collection, which can be grouped into four broad categories: marine worms, crustaceans, molluscs, and cnidarians. Burnett's work captures these encounters through what Bristow terms "active intentionality" (Bristow 2015, 50), planning the swims, but not the encounters. This results in chance meetings with a mix of endangered non-endangered species, all of them noteworthy due to the fact that they are small and usually easily startled. The unplanned nature of these encounters is crucial to providing a balanced view of swimwork, with Burnett's unbiased perspective lending an authenticity to the notion of "being in place." Her decision to explore invertebrates rather than charismatic megafauna or more popular species is vital, as "invertebrate groups are in urgent need of further assessment to pinpoint site and species-specific conservation" (Chen 2021, 2). In this sense, there is a lag in terms of awareness as well as conservation measures when it comes to marine invertebrates. Burnett highlights the vulnerability of this particularly overlooked group in the following passage: "your still being / here is a miracle … / not yet calling it off … / each of us going our separate ways together" (81). Here, membrane logic operates across both human and nonhuman bodies that have to adapt to change.

My adoption of the term hydromaterialism highlights the synergies in Burnett's writing between Neimanis's "hydrofeminism" (2012) and Alaimo's "new materialism" (2011, 2014), while focusing on the *body* as an assemblage of organised *matter*. The prefix "hydro" highlights our evolutionary aqueous heritage, as well as current material and cultural links to water. Neimanis's hydrofeminism, proposed under a wave of new feminisms where ecotones are border zones of comfort, "positions water as an ever-shifting body that connects all beings and archives all histories"

[3] My approach is closely linked to posthumanist notions advocated by Rosi Braidotti and Karen Barad.

(2012, 91), by considering the body as a body of water operating within a membrane of logic. Ecotones are part of Burnett's work, conducted at the coastal margins where different types of species congregate—a realm constantly shifting under the power of the tides. In addition to more permanent residents, intertidal ecosystems are important nursery habitats, where a wide range of animals come to lay their eggs or give birth to live young. Their offspring spend the first stage of their lives in these nutrient-rich waters, gaining strength, hardening shells, and sometimes fully changing their appearance before heading out into deeper, more dangerous waters. Neimanis uses the term "membrane" (2022, 1) to conceptualise divisions between habitats, inhabitants, bodies of water, and their surroundings. These areas are as interconnected as they are separate. Yet borders are required, and Neimanis specifies the difference between water as matter and being a body of water, the former being "materiality without meaning" (2022, 2). Therefore, matter that can be encountered and events that happen *out there*, might imply a degree of distance, but both make their way into the body through the form of toxins and plastic particles, and by accumulating there, can cause harm. Neimanis refers to membranes and the logic they impose in her early thinking about bodies of water and the amnion (2009, 163), drawing on Tuana's notion of viscous porosity, used as "a means to better understand the rich interactions between beings through which subjects are constituted out of relationality" (Tuana 2008, 188). The same "pliant weave of relations that make and remake perceptions of the plasticized world" (2019, 74) that David Farrier speaks of when referring to Evelyn Reilly's *Styrofoam* (Reilly 2009), can be encountered in Burnett's work, as it unfolds at the margins, where plastic can accumulate.

In Burnett's work this accumulation comes to life in the poem "Sea Mouse," where the speaker encounters this decomposing creature washed up on the shore. The poem focuses on the invertebrate's stomach contents:

> out of this
> worms fall soft as whispers …
>
> & out of her hair come the corpses
> of a swallowed sea (27)

Here, Burnett's metaphor can be read as a mixture of worms, expected in decaying flesh, and ingested non-degradable plastics which emerge as the sea mouse's tissues decompose, trapped in a vicious cycle of ocean currents, fragmentation, toxicity, and death. In this sense, the sea mouse's voracious hunting techniques have failed. The peeling away of soft tissues reveals a concentration of plastic, yet what is more striking is the number of particles that are not noticeable, which will be dispersed and continue to fragment into ever-smaller ones. Whether or not the ingested plastic has played a direct role in the creature's demise, it is a common occurrence in the tissues of aquatic species. There are other hints of "encounters" with ocean plastics in the collection, seen in the "nylon soil" (74) of the "Sea Anemone" poem, and in the "Common Blue" poem, where "despite the remote / location, microplastic fibres makes pools within pools" (78). This plastic entanglement is important to portray due to its destructive force in accelerating blue extinction.

Neimanis questions human perceptions of time from an immersed perspective, where "water remembers (archive), but it also forgets (dissolution)" (2022, 4). Simultaneously, water collects and dissolves other matter, facilitating an "openness for elemental feeling" (Neimanis 2022, 4). Thinking with water, in water and through water, allows Burnett to create a permeable and immersive lyric that challenges the boundaries and make-up of the human body, as can be seen in the following extract from her "Seannet" poem:

> ... as every wave
> return you to what you have always been—
> a body without end, spineless but brave
> now you're in, adrift, dripping with dream. (46)

The rhythmical "return" of the human body to its invertebrate aquatic origins connects it to the invertebrate species encountered in the present day. Transcorporeality aids this connection and can be read in the "body without end" image, which also highlights the hopeful notion of the success of evolution and preventing extinction; "a body without end" in time must be "brave" and fierce in order to survive. The imaginary captured through the word "dream" works on multiple levels: sensorial, ethical, and metaphoric. The weightlessness experienced while in the water facilitates the reconfiguring of terrestrial bodily boundaries and an openness to the same toxic elements that harm human and nonhuman bodies alike.

Burnett's practice of swimwork is a sensorial and posthuman immersion in a body of water, rendering her poetry as the ultimate exercise in lyric metamorphosis. Her submerged body in a shared hydrocommons takes us back to our evolutionary roots, a time when waters may have been less diversely inhabited but were less polluted. The feminist component of hydrofeminism is employed primarily due to new feminisms' concerns with being "open to alterity—other bodies, other ways of being and acting in the world" (Neimanis 2012, 96). Within this context, the parameters set out by a world undergoing a Sixth Extinction are shifting towards disorganised matter. The latter component of hydromaterialism is derived primarily from Alaimo's concept of new materialism, which focuses on the "material interconnections between the human body and the environment" (2011, 282). The sense of material engagement in the flows and fluxes of the world that Alaimo focuses on in her essay on *New Materialisms, Posthumanism, and Unknown Futures* (2012) reinforces the importance of the immersive element of Burnett's swimwork. In this article, as well as others published in the same timeframe, Alaimo uses the plural term *materialisms* (2011, 2014), to outline the multiple ways in which it "transgress[es] the outline of the human and consider the forces, substances, agencies, and lively beings that populate the world" (2011, 282). In the same ways in which water allows the dissolution of matter and acts as a conductive element, metaphor works in Burnett's poetry to establish more-than-human bonds between species, temporalities, and substances "in the twirling muscle / of the present / & the scream of being / one thing" (36).

Vulnerability emerges even in the hard-shelled invertebrates as early as the incipient poem, "Prickly Cockle," which is perhaps the only instance where the lyric *I* is exclusively human, and confesses:

> ... I would
> no more touch than break
> you. I would no more break
> than touch you. (13)

Via human touch, destruction seems unavoidable, yet instead of a focus on the material, in this poem the human and animal come together as *assembled matter*. There is an innate desire for discovery and contact, the connections that Burnett forges in her work and practice going beyond the physical. This builds on the notion of loss as the "bodily bringing together

[of] the different structures of nothingness—tracing their entanglements—that the world can mourn," and energetically name (Barad 2017, 56). The absence of nonhuman species and the void created by them is as important as the presence of those that the writer, or swimworker, happens to come across. These individual encounters challenge anthropocentric perspectives by broadening them to more-than-human scales where water is only one conductive element between the here and now of swimwork and the broader implications of climate destruction.

Hydromaterialism allows us to consider futurity from the perspective of a polluted hydrocommons, where water as a material will survive, circulate, contain, and be contained by non-biodegradable matter alone in a radically altered world. While Burnett does not set out to create an "anti-pure" poetics (Farrier 2019, 75), the fact that there is toxic matter in the ecosystem is inevitably captured through her practice-based methods. Identity is challenged, as can be seen in the following passage:

> All our lines break, all that's fixed shakes, when we nestle
> In the worm's wake, all our meanings fumble,
> Unable to take hold of us. (25)

The unsettled state of the swimmer's body captured in this passage ties in with Neimanis's notion of "discrete borders and boundaries" (Neimanis 2016, 46), which are challenged during swimwork. The loss of "meanings," which are "unable to take hold of us," equates to the loss and diffusion of the human subject in the aquatic environment, finding itself in a physical and cognitive sense of unravelling. Lending one's body as a tool for exploration brings with it not only a reconfiguration of hierarchies, but also a questioning of whether hierarchies exist at all. Barad sharpens this understanding of swimwork by stating that the "liquidity of pelagic habitats, alien to human understanding, may dislodge us from our entrenched approach to the world" (Barad 2017, 63). Freeing the body in this way allows for a unique perspective of the submerged body of the swimmer in the water.

Hydrocommons at the Edge

Burnett presents swimwork as simultaneously liberating and threatening, with the more-than-human body navigating between the desire for exploration and the necessity of self-preservation. The intertidal area is an entry point to an otherwise hostile environment. Wild swimming as part of this practice is proposed as a form of extinction ritual, where the act of extinction is seen as an ongoing process, the diminishing of certain species towards a permanent return to nondescript matter. Despite swimwork taking place in a hostile environment, there is beauty and wonder to be found there, sometimes in abstract forms, and sometimes in discoveries of hidden or camouflaged creatures, who transform the apparently barren intertidal landscape into a vibrant and dynamic space. Burnett's eye makes discoveries based on the capabilities of human vision, and these can be of an inorganic nature, such as plastic in all forms, shapes, and sizes, and other pollutants such as sewage-related chemicals and agricultural run-offs. The nonhuman species that live in such toxic environments remind us of the impossibility of extracting every plastic particle and pollutant that we have added to them over the years, which in some form will endure for long periods of time, prompting us to consider these global waters in renewed ways.

Burnett's approach goes beyond the observation of invertebrates' actions to questioning the reasoning behind them. This opens "a path to the recognition of shared creatureliness, whether found in the experience of enjoyment, the discernment of meaning, or the susceptibility to injury, illness, suffering, and death" (Rigby and Jones 2022, 116). This contrast can be seen where there is hope, such as in the "Silver-stiletto Fly" poem, where the fly is "keen, alive, a line, / a lilt" (30). Burnett captures the nuances of movement, direction, liveliness, and determination of the invertebrate creature in one single metaphorical and physical "line." Neimanis sharpens this thought through her statement on coastlines: "in the thickness of time, the line is actually a blur" (2022, 1). Here, Burnett celebrates the moment of the encounter and its dynamics, as she does in the "Bootlace Worm" poem, where transcorporeality is encapsulated, "[b]ody within body, whirligig" (36). Elsewhere, a brimstone butterfly confesses: "I take my body / to the margins. / Hold. Hover" (54). This act of placing oneself peripherally is echoed in the general exclusion of invertebrates from extinction discourse. There is a particular attention to stratification here: the sections titled "Below," "Above," "Surfacing," and "Spilling," resonate with the opening poem of *Swims* (2017). This poem gives the poet's definition of the verb "to swim," and contains this trans-corporeal passage:

To be at one
in the body
and under
and over it. (13)

Burnett's approach is more ambitious in *Of Sea*, which builds on her experience of writing *Swims* (2017) and shows her honing her observational and creative skills in capturing these encounters through her writing. In contrast to Tom Bristow's view that Thomas Kinsella's work encourages "being with things in our dwelling places" (Bristow 2015, 50), Burnett encourages the reader to engage with the more-than-human outside of comfort zones and habitats in open water. The text might mislead the reader into thinking that there is an assumed dichotomy between the encounters with invertebrates and the vertebrate body of the writer; however, Burnett's decision to look in that direction is deliberate. Extinction discourses and those centred on endangered species often fail to address the fact that these losses and risks are also happening to the smallest and most inconspicuous of species.

What Neimanis terms "membrane logic," whereby we "require membranes to keep us from total dissolution" (Neimanis 2022, 1), can help facilitate connections to these species, particularly from the immersed perspective that swimwork entails. Burnett singles herself out in this instance as "the only human body fibres identified in deep-sea water" (78). The word "fibres" implies solidity, yet it does not exclude fragmentation. Plastic and organic particles eventually break up into smaller, less-visible ones, with nano-plastics being invisible to the human eye. Burnett acknowledges this multi-faceted layering that occurs to creatures in aquatic habitats, as can be seen in the following extract from "Sea Anemone":

she
asphalt soil
nylon soil
sandy soil (74)

The gendered sea anemone highlights both natural and artificial substrates, while hinting at decomposition as well as the Anthropocene itself, identified through human-made changes that can be seen in the geologic record. Permeability and a return to water, coupled with the dissolve in the passage above, show not just where the body resides, but what it contains—a filter for both man-made and naturally occurring substances. The layering observed by Burnett showcases the geologic changes that are to be found in nature, as well as the anthropocentric mark-making that comes with this supremacy. Read in this vertical way, the extract shows how humanity tramples on the delicate body of the invertebrate creature, paving the way for a new range of networks that seemingly benefit the human, but run the risk of bouncing back and harming human and nonhuman creatures alike.

Scales of Blue Extinction and Deep Time

Burnett's swimwork highlights the importance of adapting to various scales and timeframes in order to consider the full implications of blue extinction. Zooming in and out during practice-based work enables the writer to capture minutiae within the broader sphere of climate destruction. The problematics of water Burnett identifies in *Of Sea* extends to the hydrosphere. Throughout *Of Sea*, Burnett establishes connections between the human *I* in their immediate observations and the more distant elements beyond the horizon. In "Sand Crab," the lyric *I* is pulled out to sea and to the zoomed-out perspective afforded by open water:

> I balance my current only takes me back
> when seawards seawards is the call of my curve (63)

The word "seawards" is repeated in a transcorporeal chant, where the more-than-human body is at the mercy of the energy, flows, and shifts of water that take hold and pull it out to sea, working against "meaning" (25). The immersive experience requires "balance," as does the environment that Burnett explores. Drawing on Thom van Dooren's views on what he terms worlds of meaning at the edge of extinction (2023), the choice of location and the value of swimwork become apparent.

Addressing marginalia in her practice-based writing by inhabiting often overlooked blue places allows Burnett to consolidate van Dooren's worlds of meaning at the edge of extinction, intervening "in these processes, practices, and relationships in order to create new possibilities for shared life in diverse more-than-human communities" (van Dooren 2023, 6). Burnett's proximity to marginal marine invertebrates is key in triggering responses in Western audiences. Her swimwork operates in a type of micro-sacrifice zone, an idea derived from David Farrier's theory of "oceanic sacrifice zones" (Farrier 2019, 11). Readers are therefore made aware of aquatic areas that are excluded from mainstream conservation narratives. This was one of the key issues identified in the *Extinction Stories Matter* paper when it comes to the ability of literature to trigger emotional responses in readers, which might result in pro-environmental attitudes and actions (Malecki, Wiek von Mossner, and Sorokowski 2023). Contrasting with other texts where the spotlight is on species who are usually found in the Global South and are too exotic for a Western readership, *Of Sea* looks for inspiration on its own doorstep.

While Burnett's focus lies on invertebrates rather than larger species, her writing extends the issue of scale when it comes to extinction and vulnerability. Large animals may attract attention more easily than smaller ones, which require an effort, an attentiveness, a moving into the space that the animal inhabits. The process of blue extinction for the most part lacks this component due to the remoteness and inaccessibility of these environments. Burnett intervenes at the periphery, fine-tunes herself to the small, the almost invisible, as can be seen in the "Sand Flea" poem, where the invertebrate is "missed. The size of the space / between words [...]" (51). The size marker is simultaneously physical and an extended metaphor for scale and speech. Small invertebrate creatures run the risk of falling between the cracks of conservation narratives and yet play important roles in intertidal ecosystems which are primarily nursery habitats for a large number of species due to high nutrient levels and the relative safety offered by shallow waters.

Rebecca Giggs focuses on the issues of scale from the perspective of large-bodied marine creatures in her book *Fathoms: The World in the Whale* (Giggs 2021, 11). While whales come with a myriad of issues of their own, primarily due to their commercial desirability, living in waters shared by intertidal invertebrates renders them subject to the same risks and effects of pollution. In Burnett's "Barrel Jellyfish" poem, the collective "we" asks the reader to "[l]ook for us, even after we are gone" (16), highlighting the imminent loss triggered by these current threats. Yet hope and the quest for identity remain; both on a physical and metaphorical level, the search continues. The plural "we" emphasises the issue of scale through a multiplication of the lyric "I" and the implied extension to the more-than-human in its employment. Giggs also observes this in the following passage: "We are all tumbled together, human and nonhuman, the far and the nearby, deeply in torsion, inhabiting this change of state" (2021, 250). In "Barrel Jellyfish" Burnett also states: "We resemble" (16), with these remarks drawing on transcorporeality as a "commonality that still insists on difference" (Neimanis 2016, 38). Place-time scales are extended from the individual human body to the mass of more-than-human bodies with which the Earth's waters are shared. Within Burnett's lyric, metaphor fluidly facilitates this shift and is in line with Farrier's view that it "produces a new 'knot' or cluster of understanding—a resolution of some form—without losing a sense of abiding difference" (Farrier 2019, 97). The section from "Barrel Jellyfish," "[w]e put our parts / together. We outward / form" (16), reminds the reader of the membranes at play in this physical and metaphorical coming-together. Neimanis highlights the fact that the vital role of water in "[t]he passage from body of water to body of water is never merely metaphoric, but rather radically material" (2009, 164). Yet in Burnett's work metaphor and material combine to create a lyric membrane logic underpinned by the body of water in water and the writing that emerges from this experience.

The micro-macro lens that Burnett adopts in her writing practice is deployed in a place where the layer of water is peeled back constantly under tidal rhythms. In "Barrel Jellyfish" Burnett creates a visceral metaphor and hypothesis: "[s]ay we were whale fat" (16). In doing so, she selects the largest commodity obtained through whale hunting from its blubbery body, which is widely used to this day, and which is also a storeroom for toxic chemicals and pollutants. The repository afforded by the whale's body as metaphor works beyond its eclectic amalgam of gathered toxins, adding a new dimension to Giggs's question: "How much is any whale, or any animal really, an inadvertent ark or cultural archive of things we have lost touch with?" (Giggs 2021, 126). Reconnecting the body to other bodies is reinforced through Burnett's swimwork, where the process of discovery and arbitrariness is also a process of de-archiving blue knowledge that permeates more-than-human membranes. Within the micro and macro scales applied in *Of Sea*, "[t]hese bodies are all caught up in one another's currents—as they are with the whale's body, the body of the rain cloud, and the body of the increasingly toxic sea" (Neimanis 2016, 38). In his reading of Mark Doty's poem "Difference," Farrier reflects on the role of the more-than-human body captured in the text and culminating in a form where "[e]ndlessly malleable, the jellyfish becomes simply 'the stuff / of metaphor'" (Farrier 2019, 97). For Burnett, the barrel jellyfish represents a physical body as well as a metaphorical one, which is then held within thin rectangular membranes of ink on the page. Enjambment works across these linear frames to highlight the permeability of language, metaphor, and matter. Continuing her exploration of alternative matter, Burnett states "[s]ay we / were good for the economy, or don't say anything" (16). Ending on silence in this passage becomes equivalent to the problem of not speaking up for the more-than-human. The voice that travels through the membrane of silence belongs to creatures who cannot speak or lack mouths altogether.

Whales swim into shallow waters for two main reasons: to calf and to strand—for life and death, two contrasting events driven by cyclicality and confusion. The rhythms and traditions present in whale communities and whale pods are broken when stranding events occur. There has been a global increase in these horrifying events, and while the reasons behind them are complex, whale bodies have been found not only to have copious amount of plastic in them, including an entire greenhouse on one occasion, but also developmental issues linked to toxins in their bodies (Giggs 2021, 10). Underdeveloped sexual organs point to reproductive risks, while emaciation highlights depleting feedstocks (van der Hoop, Corkeron, and Moore 2017). Fish that whales normally eat are now harvested either for human consumption or industrial use. Baleen whales are filter-feeders, and like the majority of species that Burnett focuses on in *Of Sea*, they spend their time sifting through water and capturing both nutritious and toxic particles.

Water does a spectacular job at concealing such large-scale issues until they come to light such as when they are washed ashore; however, if this subaquatic layer were peeled back, it would make a lot of these realities impossible to avoid. For example, coral reefs are portrayed as a large entity, but they are an assemblage, and even if viewers zoom in to the level of an individual coral, they are not necessarily observing the individual coral polyp. Coral reefs highlight one of the key problematics of aquatic extinction, which is that it remains predominantly unseen, concealed by the water. If coral reefs were to die out on land, the putrid smell they would release would be too strong to ignore (Gillespie 2016). Van Dooren and Giggs demonstrate the importance of actual encounters to the apprehension of extinction in its full-bodied reality. However, if the world population were to seek direct encounters, it would quickly turn into an ambush against wild species, and the ecological consequences would be devastating. Relying on the power of literature to capture these encounters and share them with readers is therefore vital for efforts to conserve blue ecosystems.

Burnett's *Of Sea* is remarkable in the way that it details invertebrate encounters at a time of environmental crisis and under the threat of extinction. This collection provides ample ground for exploring hydromaterialism and its mechanisms, through polluted, absent, or dynamic waters, and via the immersion of the writer's own body in these places that other creatures call home. Despite differences in the method, structure, and approach of the poems, Burnett offers the reader the same embodied experience of place throughout, finely filtrated to seemingly imperceptible encounters and deploying metaphor to redefine the place of the more-than-human body in a dynamic and interrelated world. Drawing on the concept of membrane logic allows for a poetics of place and a poetics of water, where hierarchies and boundaries mix and dissolve. The swims are planned, but the encounters are unforeseen and often surprising. Burnett's experimentation with shifting micro and macro scales offers a sense of locality and immediacy of the encounters, coupled with a wider consideration of blue extinction as an event which is unfolding on a global scale, submerged yet all subsuming. The sense of multispecies kinship expressed throughout the text likewise offers hope for the future and a more constructive relationship with aquatic invertebrates. Burnett's embodied practice implicitly prompts a better understanding of aquatic environments and the conservation methods that can be applied to protect them and prevent further destruction.

References

Alaimo, Stacy. 2008. Trans-corporeal Feminisms and the Ethical Space of Nature. In *Material Feminisms*, ed. Stacy Alaimo and Susan Hekman, 237–264. Bloomington, Indiana: Indiana University Press.

———. 2011. New Materialisms, Old Humanisms, or, Following the Submersible. *NORA—Nordic Journal of Feminist and Gender Research* 19 (4): 280–284. https://doi.org/10.1080/08038740.2011.618812.

———. 2012. Sustainable This, Sustainable That: New Materialisms, Posthumanism, and Unknown Futures. *PMLA* 127 (3): 558–564.

———. 2014. Oceanic Origins, Plastic Activism, and New Materialism at Sea. In *Material Ecocriticism*, ed. Serenella Iovino and Serpil Oppermann, 111–142. Bloomington: Indiana University Press.

———. 2017. Your Shell on Acid: Material Immersion, Anthropocene Dissolves. In *Anthropocene Feminism*, ed. Richard Grusin, 89–120. Minneapolis, Minnesota: University of Minnesota Press.

Barad, Karen. 2017. Troubling Time/s and Ecologies of Nothingness: Re-turning, Re-membering, and Facing the Incalculable. *New Formations* 22: 56–86. https://doi.org/10.3898/NEWF:92.05.2017.

Bristow, Tom. 2015. *The Anthropocene Lyric: An Affective Geography of Poetry, Person, Place*. Basingstoke: Palgrave Macmillan.

Burnett, Elizabeth-Jane. 2017. *Swims*. London: Penned in the Margins.

———. 2021. *Of Sea*. London: Penned in the Margins.

Chen, Emily Y.S. 2021. Often Overlooked: Understanding and Meeting the Current Challenges of Marine Invertebrate Conservation. *Frontiers in Marine Science* 8: 690704. https://doi.org/10.3389/fmars.2021.690704.

Deakin, Roger. 1999. *Waterlog: A Swimmer's Journey Through Britain*. London: Chatto & Windus.

van Dooren, Thom. 2023. Worlds of Meaning at the Edge of Extinction: Conservation Behaviour and the Environmental Humanities. *humanities* 12 (122). https://doi.org/10.3390/h12050122.

Farrier, David. 2019. *Anthropocene Poetics: Deep Time, Sacrifice Zones, and Extinction*. Minneapolis, London: University of Minnesota Press.

Giggs, Rebecca. 2021. *Fathoms: The World in the Whale*. Melbourne and London: Scribe UK.

Gillespie, Katherine. 2016. Turns Out Dying Coral Reefs Smell Even Worse Than They Look. *Vice*. June 7. https://www.vice.com/en/article/xdm8a4/diving-in-the-stinking-decaying-great-barrier-reef. Accessed 5 April 2024.

van der Hoop, Julie, Peter Corkeron, and Michael Moore. 2017. Entanglement is a Costly Life-History Stage in Large Whales. *Ecology and Evolution* 7: 92–106. https://doi.org/10.1002/ece3.2615.

Malecki, W.P., Alexa Wiek von Mossner, and Piotr Sorokowski. 2023. Extinction Stories Matter: The Impact of Narrative Representations of Endangered Species Across Media. *ISLE. Interdisciplinary Studies in Literature and Environment* 30 (4): 1–19. https://doi.org/10.1093/isle/isab094.

Neimanis, Astrida. 2009. Bodies of Water, Human Rights and the Hydrocommons. *TOPIA* 21: 161–182. https://doi.org/10.3138/topia.21.161.

———. 2012. Hydrofeminism: Or, On Becoming a Body of Water. In *Undutiful Daughters: New Directions in Feminist Thought and Practice*, ed. Henriette Gunkel, Chrysanthi Nigianni, and Fanny Söderbäck, 96–115. New York: Palgrave Macmillan.

———. 2016. *Bodies of Water: Posthuman Feminist Phenomenology*. London: Bloomsbury Academic.

———. 2022. Hydrofeminism on the Coastline: An Interview with Astrida Neimanis, ed. Sarah Bezan. *Anthropocenes – Human, Inhuman, Posthuman* 3 (1): 8. https://doi.org/10.16997/ahip.1363.

Reilly, Evelyn. 2009. *Styrofoam*. Roof Books.

Rigby, Kate, and Owain Jones. 2022. Roadkill: Multispecies Mobility and Everyday Ecocide. In *KIN: Thinking with Deborah Bird Rose*, ed. Thom van Dooren and Matthew Chrulew, 112–134. Durham and London: Duke University Press.

Tuana, Nancy. 2008. Viscous Porosity: Witnessing Katrina. In *Material Feminisms*, ed. Stacy Alaimo and Susan Hekman, 188–213. Bloomington: Indiana University Press.

Whitworth, Victoria. 2017. *Swimming with Seals*. London: Head of Zeus Ltd.

CHAPTER 9

A Story of Eight Limbs

Rosanne van der Voet

> *We should not think that its symbolic stories reduce the sea to a cultural signifier; rather, these stories are products of an imaginative impulse to visualize through poetic language terrestrial-aquatic interactions and encounters. So, the stories of the sea matter as much as its materiality, and despite creating a problematizing tension in the sea discourse, these qualities are coextensive, provoking an investigation of how marine meanings are crafted and lived in a palpable sense between the natural and the cultural, and how they affect perceptual transformations, ideological shifts, and amendments in knowledge production.* (Oppermann 2019, 450)

R. van der Voet (✉)
Leiden University, Leiden, Netherlands
e-mail: r.van.der.voet@hum.leidenuniv.nl

© The Author(s), under exclusive license to Springer Nature Switzerland AG 2024

V. Fibisan, R. Murray (eds.), *Blue Extinction in Literature, Art, and Culture*, Palgrave Studies in Animals and Literature,
https://doi.org/10.1007/978-3-031-69910-8_9

HEAD

Soft body slowly flows into head, stomach and hearts drifting behind brain. A spherical sac buoys body, blood and organs held in place. Face in front, eyes emerge atop a rounded structure, swelling into saltwater, intercepting shape-shifting skins hiding under rocks. The white of eyes centres around an elongated black pupil, window through which a mind looks out. Light approaches, mind thinks up a visual overview of surroundings. Clasping itself onto throat, brain trembles slightly, moves with body as a beak scrapes a soft fleshy being out of its shell. Ink sac between gills shivers with expectation as body retreats, pushing into crevice. Watch a current flow, the smallest movement of a neighbour in an adjacent den. Eyes survey the rocky outcrop thick with movement, an anomaly in the flat, outstretched sandy seabed. Debris, collected, lies scattered across the sand. Little mounds have formed, miniature territories within shared space. More objects encroach on den, remains of others' meals. They are expelled with a fling of front arm acting independently, the embrained limb shooting the calcareous skeletons away, now drifting downward at the edge of the rocks. Eyes watch a small sand cloud appear where they land.

Face looks into underwater world, they say gloomily. Yet no human-given names are at the forefront of mind. They say this creature is unexpected, part of an anomalous community of unusually social specimens (Scheel et al. 2017, 291). Body is imagined as being solitary, its gloomy facial expression foreshadowing its solitude. Yet more central is the meaning of a missing shell, a misplaced crab leg, a gleaming beer bottle stuck in the sand, radiating green. Eye rests on a tentacle twitching, rapidly changing colour. All these things unfold towards body as mind gives them meaning, as eight-limbed consciousness becomes aware of them. Multi-limbed mind categorises aquatic dust particles drifting just out of reach, their meaning when they come closer, the gritty feeling on suckers. Collapsing together, expressively commingling with one another, the sum of this being and others forms a narrative, a voice (Oppermann 2019, 459). A thought implores brain, nudges curiosity toward neighbour's arm, now purple, yellow, blue, smooth at first, then spiky. Arm extends to the fresh pink suckers, delicately curling up. Body longs to feel the softness, the elastic tissue enveloping elastic limb. The skin parts at touch, the stickiness of another's body. Gently, arm touches arm, strokes the shapeshifting skin, suckers on suckers. Arm and body sleep on, gills slowly breathing water. Colourful dreams reflected on skin send out octopus stories in patterns of colour, exhibiting themselves to the sea around.

Despite

so many
limb-brains

ecocriticism's

ocean deficit

effaces this
sticky
invertebrate
arm

*Western
literary
history*

*(as well as
global
literatures)*

is awash

limbs
resurfacing
from negated
depths

*with
representations
of ocean.*

Seas of ink

flow out of
this body in
protest

spilled

and
spill

*in oceanic
adoration*

in chemically
altered
saltwater
darkening

multiple
selves hover
for a moment
outside their
camouflaged
interior

and

despair.

(Dobrin 2021, 15)

By expanding on the models and ideas of marine sciences in creative ways, material ecocriticism asks, What if cognitive aesthetics is reconceived as the 'cognitive vibrancy' of sea creatures that exceeds the all-too human frames? What if there are narrative pathways for marine organisms to express themselves in contingent patterns of creativity? (Oppermann 2019, 451–452)

I.

arm by arm

this story composes itself

curling through grounded narratives

all-too-human frames

interweaving them into what flows and sticks

diffusing in sepia secretion

into the stories

narrative pathways

inscribed on oceanic flesh

the smallest movement of a limb

holds a multitude of parallel stories

that tell themselves

inside of me them she

simultaneously

voices coalescing

in each saltwater drop

Stories are not just told in black letters against a white background. They flip their tails, dive into acidified waters, bob in plastic soup, churn in the oceanic gyre. They pulse their fins, stretch human notions of time in old and slow shark-bodies. A story erupts in a suddenly spiky and violet arm, soft and yellow only moments ago. In a shapeshifting skin cell, a reaching limb swimming into a human story, translated, questioned, misrepresented. A story precedes storying, unknowable, nonhuman. Why should these words manipulate them and present their meaning? This arm is now a narrative pathway, her eye is a signifier, the beer bottle symbolic of this fish-bowl vision. A wave is imagination, a sea cucumber's knowledge in your tube feet tiptoeing the sand erupting from her headless limbless body. These human words bestowed your meanings, yet somewhere tremble the stories unmitigated by a human brain and artificial eye. Do they want to be told in this language? What happens if these words disregard them?

And so uncertainty overflows this page, saturates it in saltwater. This language is always problematic, limited, presenting merely a shred of a story. Yet what happens when the vibrancy of ocean stories overwhelms this human impulse, the course of this sentence? When words are mobilised against a forgetting of ocean and destructive human impacts? When unknowability immerses these words and allows for a drifting in the storied ocean? For a trembling of human hands, a hesitation in a finger's descent to keyboard, letter, word, meaning. When a human-written story dissolves human truths, immerses itself in its contradictions, the different knowledge of ocean stories. When species' humility makes readable the spectrum of colours in the oceans. The flowing backscatter of plankton and small creatures compromising the uniformity of the backdrop, background becomes a substance not surface. Immersed and ever trembling, ocean stories are told from an ecosystem-embedded position, narratives inseparable from bodies of saltwater. The oceanic world is full of living connections, each telling a small story, together forming an ever-expanding chain of narratives. When scientists observe a group of gloomy octopi living in close proximity to one another, exhibiting complex social behaviours, this is described as unusual, unexpected. Yet such communities can also be seen as 'the narrative power of bustling, coexisting life forms,' extending their own specific ways into their watery worlds, asserting their selfhood (Oppermann 2019, 460). No matter how they are described, the ocean and its inhabitants keep emerging, finding their way into human consideration through technological mediation, unsettling terrestrial, anthropocentric and grounded forms of knowledge. As Serpil Oppermann points out in her work on the narrative agency of the ocean, the sea,

> literally and figuratively, wants us to be attentive to its meaningfully articulate inhabitants [...] such attentiveness takes the storied sea, its physicality detailed in scientific research, and their dialectical relation seriously. The storied sea today is a hybridizing mix of the Anthropocene dilemmas within which marine creatures play out entwined ecological crises and material intimacies. And, whether they live in the pelagic or benthic zone (where no sunlight penetrates), they want their voices heard, their stories recognized, and their attempts to stay alive understood. (Oppermann 2019, 452–453)

To know the storied planet, drones and satellites transcend oceans and continents. Composing a god's eye view in blue and green, all-encompassing images are formed, revealing the oceans as a blue blanket covering the earth's surface. However, the 'view of the earth from space reveals merely the surface of the seas, a vast horizontal expanse that is rendered utterly negligible when one considers the unfathomable depths and three-dimensional volume of the rest of the ocean' (Alaimo 2016, 161). Such images erase depths, heights and other subtleties. They do not ask which creatures dwell on sandy seabeds hundreds of metres beneath that static ocean surface and which narratives are told in their movements. They do not ask what stories can be read in eight soft arms gently curling, in cephalopods living in dense congregations. For when human eyes and human-made cameras descend into oceanic depths, any sense of overview is lost, replaced by images of underwater murkiness, of the occasional strange limb making an unfamiliar movement (Alaimo 2013, 235). Yet descending is the only way to know ocean depth, to uncover a glimpse of underwater stories. To begin to reveal ocean narratives, 'one must descend, rather than transcend, be immersed in highly mediated environments that suggest the entanglements of knowledge, science, economics, and power' (Alaimo 2016, 161). And so these words sink through digital spaces into open oceans, mingling with saltwater and inky secretions. Slowly diffusing their human meaning, they spill into the margins of this text, swelling with strange and unexpected intimacies.

descending is becoming immersed

to be drawn in

arms coiling

to listen to all at once

some are subtle

first arm

heralds

around arms

some

doubtful

soft

sticky

extends

sticky voices melted

into scattered

self

II.

a multitude of eight-armed beings
reconstructs a rocky outcrop
'ecosystem engineers' (Scheel et al. 2017, 286)
excavating
exhibiting
altering its suitability for other species

ironically
lead fishing lures distinguish collection
a camera descends
body surges upward
elongating itself in alarm
sharply pointed hood floating tall
an intrusive octopus is evicted

look into gleaming red-black surface
which is a depth
a second arm reaches tentatively
suckers feel the smooth material
its lack of fleshiness
some part of this dark eye
materially enters this being (Alaimo 2013, 236)
mediated arm reaches through it
emerges on a screen
enters human retina

creating 'a narrative emergence' (Oppermann 2019, 454)
when human stand-ins
'dive into the ocean to unveil the messages
hidden among agentic entities'

What possibilities does this eerie and entrancing hue pose for new materialist and posthumanist ecologies of the depths? And how do the prismatic bioluminescent displays of creatures in the abyss provoke recognitions of the multitude of aquatic modes of being, communicating, and knowing? Violet-black ecologies of the abyssal zones entice us to descend, rather than transcend, to unmoor ourselves from terrestrial and humanist presumptions, as sunlight, air, and horizons disappear, replaced by dark liquid expanses and the flashing spectrum of light produced by abyssal creatures. (Alaimo 2013, 235)

III.

Descending, familiar hues slowly disappear. Red goes first, swiftly followed by orange, yellow, finally by green and blue (Alaimo 2013, 236). Gloomy arms coiling towards each other from adjacent dens do not extend to these zones of deep dark. Yet other eight-limbed cephalopods swim within the abyssal violet-black, which not only absorbs all light, but also swallows terrestrial human knowledge. For here life does not depend on sunlight, which is often quoted as conditional to all life forms (Alaimo 2013, 245). In these depths, dread of the unseen abyss engulfs the human mind, materialising directly in the endless darkness beyond the small beam of a searchlight. Turn around, swirl in any direction, and the black envelops all. A mirage of violet seems to drift at the edge of the darkness, in the corner of eye, a certain roundness confining the black. Yet when looking straight at it, any sense of edge is reabsorbed. Violet-black enshrouds body, becomes a viscous substance that slowly swallows everything, out of which there is no way back to air, sunlight, land. All around drifts the unseen, it hovers at fins and fingertips, it flows on skins and scales. A thin beam of light accentuates its monstrosity—the light illuminates nothing, reinforcing the dread of what lies just beyond (Dobrin 2021, 198).

Colour at this scale cannot be articulated, the limits of eyesight determining the limits of mind. Barely aware of the specificity of a four-limbed body, of a single big brain, unwebbed digits and sunlight-dependent perception. Eye approaches the unseen the same way it approaches the seen, unwittingly straining to catch a glimpse of colour. Violet-black overwhelms senses with its unrelenting nonhumanness, its lack of recognition of perceived human importance. Too slowly, it bears down on this body, painfully, until disintegration starts. It burdens a human body in the way speciesism sanctions the violent capture of eight-limbed oceanic beings. The trapping of a soft body in a cage, chains tightening around throat and brain. Hauled up, so many seas pass by this face, a rushing of saltwater tingles skin. Exposed to air, soft flesh tightens and contracts in suffocation as body is cut into pieces alive. As arms are severed, consciousness disintegrates, far too slowly, extinguishing in excruciating awareness of its demise. Arm by arm, self is scattered in agony, finally evaporating to emerge a slimy mess on a plate with arms sometimes still wriggling, suckers still trembling the last shreds of this octopus story.

The violet-black does not absorb, it enters body, materially penetrates it, completes the enmeshment of surroundings and flesh. Like a suffocated octopus, a human body drifts in the abyssal zone, without say over its survival. This zone, constituting 78.5% of all earthly habitat, marginalises human modes of perception, effortlessly swallowing anthropocentrism and terrestrial frames of reference (Alaimo 2013, 234, 236). In this violet-black ecology, the human body becomes the alien, hovering at the limits of the abyssal ecosystem's comprehension (Alaimo 2012, 477). Blackness is not the same as nothingness, and descending does not mean going back in time: 'The deep seas, once thought to house "living fossils" that terrestrial time left behind, are in fact home to sea creatures who live at a slower pace, within the cold, dark, and heavy waters' (Alaimo 2016, 161). In the violet-black depths, the quality of being eight-limbed is stretched and condensed into all kinds of appearances. The abyssal cephalopods are not necessarily ancient, possessing an evolutionary history just like terrestrial animals, always in the process of adapting and evolving.

Many deep-sea dwellers remain unrepresented in human systems, uncaptured by human technologies descending into seas. Of some of them, merely a few blurry images exist, as is the case for these spiderlike limbs, ghostly white against the black sea beyond. Arms extending up to eight metres into the sea around, elastic limbs pulse and flow behind mantle. Body and head obscured by large fins, the centre of mind is lost in hazy recordings made with underwater cameras. Arms start in a straight sideways line, then dropping for metres down beyond, perpendicularly, they say elbow-like. Time flows slowly like body, awaiting peers in the white nebula of head. Sometimes body drifts on undersea currents, matching the arrival of zooplankton to dance into its centre. Orifice leads to a squid's digestive tract, hearts pound behind it a deep blue blood. Through body, through the whiteness of limbs sometimes pink, a deepened complexion is lost on imagery in which blood does not flow, body does not pulse. Soft arms sometimes caress the seabed, gently lift up a polychaete body. Mouth swallows others and awaits the next drift. Arms are for reaching, for pulsing in a certain direction. They are for stretching towards things that are not this body, a plankton cloud, a worm, an uncomprehending dark eye descended. Arms interweave this flesh with other flesh, with saltwater always filled with the smallest bodies. Arms stretch mind, reach out beyond. These limbs are the potentiality of body as it grasps a bioluminescent message of danger materialising, as it speeds to further depths. They are the mark remaining for a moment when this body has left the scene.

Represented in black letters against a white background, or as a white body against a backdrop of black sea, this animal is more easily imagined with eight short arms joined by umbrella-tissue. It is preferred with two prominent earlike fins, a small silver body gently drifting and flapping. A body unspiderlike and comforting, a living embodiment of human nostalgia. That does not make it Dumbo, but this appearance appeals in uncontrollable ways. One of the deepest dwelling octopi, not much is known about this being. Yet this small body, unusual fins and far-apart eyes remind of a famous fictional elephant. Elephant-morphised and anthropomorphised, this octopus is taxonomised as a cultural icon. A scientist bestowed this most nostalgic name, a metaphor for a character who is a metaphor, connected to so many things, so many human meanings

rippling

and rippling

and rippling out

what this body means

is obscured by its alleged cuteness
by a specimen collected

hauled up
transcended
dissected
imploded
assembled
and stored
and stored

yet this self

drifting through abyssal sea
embrained limbs holding sense
in small beads of lucidity

an arm swings
pouncing onto amphipod body
swallowing crustacean whole

polychaete patterns in the sand
weight of oceans on skin
shadow of goblin shark

nothing unmakes this knowledge

it is there as it has always been

unrelenting

there is no softness in these eyes

no humanity

this ipseity proves forceful

if enticed to descend instead of transcend

and descend

sinking words

drifting into abyssal seas
conceiving a violet-black ecology
in highly mediated ways

reach zones of darkness

to witness the eight-limbed
constructing worlds

IV.

Though devoid of sunlight, the violet-black world brims with untold lights. Bioluminescent signals allow for communication in darkness, extending a language of light into violet-black waters (Alaimo 2013, 246). Sometimes there is a faint white glow in the distance, or a flickering right nearby. At other moments overwhelmingly bright and colourful flashes engulf the black. Bright hues illuminate small creatures within each saltwater drop, drifters on undersea currents. The faint backscatter, an oceanic drizzle, constitutes the collective breath of an ecosystem ebbing and flowing in shimmers of light. Immersing the human eye in bioluminescent cultures via technological mediation undermines the idea of the ocean void, inspiring the knowledge of the abyssal zone as a vibrant ecosystem. A posthuman descent into the deep sea takes on an unusual form. It is not a net scraping seabeds in search of specimens for classification, nor a machine crashing into rocks and bodies. Instead it is a camera beyond perceptible light frequencies entering the deep sea unnoticed, recording blinking lights of alarm, seduction and distraction, observing unknown semiotic systems. It is an electronic jellyfish plunging into abyssal seas, lighting up, provoking bioluminescent displays (Alaimo 2013, 246). It is a prosthetic eye, a jelly-robot slowly learning deep sea languages of light, gradually recording the meanings of a red flash, a green-blue glow, a sudden glare. A jelly-robot's tentacle, gently extending toward cephalopod arms, joining flesh with machine. It is jelly-robot data being interpreted by a human mind unmoored of grounded, terrestrial frames of reference. A human mind adrift in violet-black, opening itself to bioluminescent deep sea knowledge. To an anglerfish's lure, a sea firefly's seductive display. To a shining tubeshoulder squirting light, a sea cucumber exhaling a pink glow from its transparent body. To an octopus's suckers twinkling blue and green *l*

a

d

e

n

p

r

o

m

i

s

e

o

f

t

h

i

s

b

o

d

y

While the human alterations of the geophysical landmasses of the planet can be portrayed as a spectacle, the warming and acidifying oceans, like the atmospheric levels of CO2, cannot be directly portrayed in images but must be scientifically captured and creatively depicted. The depths of the ocean resist flat terrestrial maps that position humans as disengaged spectators. Marine scientists must, through modes of mediation, become submerged, even as persistent Western models of objectivity and mastery pull in the opposite direction. The substance of the water itself insists on submersion, not separation. […] The oceans proffer a sense of the planet as a place where multiple species live as part of their material environs. As human activities change the chemical composition, the temperature, and the alkalinity of the waters, marine creatures also change. (Alaimo 2016, 161)

V.

As carbon dioxide is released into the air, carbon is exchanged between the atmosphere, the oceans and land. Reacting with seawater, carbon dioxide forms carbonic acid, which swiftly dissolves to form bicarbonate ions and hydrogen ions. Particularly in colder oceans, which absorb more carbon, the increased concentrations of hydrogen ions create a higher acidity of seawater (Evans Ogden 2013, 322–323).

this fifth arm trembles

In more acidic waters, creatures with calcareous skeletons are threatened,

slowly

disappearing

They rely on absorbing carbonate from water, the acid now

permeating skin disintegrating self

beginning to dissolve their bodies,

the skeletons and shells which hold them in place (Evans Ogden 2013, 323).

from centre this arm

grips onto coral

its p o r o s i t y

makes it so the slightest pressure

c r u s h e s it

a clownfish foolishly approaches

traces shadow as body shoots through water

small cloud of dust descends onto sand

peripheral vision blurring

ocular development stalled by lack of oxygen (Dobrin 2021, 177)

faint amber

out it swims

out

away from reef into dark blue unknown

there it goes

small orange speck in dark blue

this Nemo *will not be found*

its sense of smell no longer capable
of recognising the right habitat
of distinguishing a potential predator
from its own kin (Evans Ogden 2013, 324)

the real fish who does not speak English
the fish expelled from life
due to its acidity-impaired perception

who will drift

ironically and unlamented

remaining forever unfound

unrepresented
extinct
and materially depths and depths
of saltwater away

from any human representation of it

as Nemo goes

food sources are absorbed in blurry outlines
the reaching of an arm no longer suffices

clownfish, shells and corals are replaced
by yet more amphipods and tanaids
collective biomass is reduced (Evans Ogden 2013, 325)

arms

hurt

starved

contract

and retract into body

VI.

A bright pair of eyes watches vision dissolving on acid currents. Fish have flown, octopi are starving and exoskeletons are dissolving. Still this creature sits here, body faintly swelling and dropping, exhaling water, oxygen, there is some left, some. A faint glowing far away, the sun above the surface of the water, will eye ever see it. Will body float to the surface and stare into it, counterintuitively, blinding eyes forever with a beautiful last shimmer. The sunlight that permeates the coastal seabed coats body in a thin green-blue glow. Body is in deep seas and pelagic zones. It is in coral reefs and coastal bays with rocky outcrops. These eight arms stretch out into all oceans of the world, taking on a myriad different forms, bearing witness to countless different extinctions. Resulting in an emptiness, a simplification of what used to be. This body, in its current capacity, is no longer to be, no longer. To float no longer, no longer to rush through kelp, for skin to shift and discolour. No longer for this story to be obscured by an octopus ink cloud, for its meaning to be disguised like an octopus presenting as a pile of shells. No longer for these words to belie their own semantics, to worm themselves unexpectedly into crevices. Into the margins of this sea, this sea of words, exhibiting its doubtful certainties. No longer for dreams to be reflected on skin, for cephalopod stories to emerge and bustle in manifestation of a strange narrative agency. In the Anthropocene, eight-armed stories are needed and lost, coiling tentacularly for just a shred, a shred, one arm, cut off if need be, to survive and to tell. A severed limb grows back but it is no longer the same, does not tell the same story.

This sixth arm coils, extends, forcing itself out of body. Gripping onto a whelk, just a common one. Yet there never was a whelk quite like this, undone of its fleshiness, slowly dissipating into history. Holding it close to face, see the intricate pattern spiralling inward and outward. A process of composition seemingly infinite, and yet the result is held in these arms. The shell perfectly embodies the illusion of separation, the rift between perception and how things are. Now shell spirals in suckers, twisting and spinning, its surface thinning, becoming ever more transparent. It floats in front of face, slowly every molecule dissolving in acid. The spiralling shell enthrals this mind, makes the boundaries between self and other

things c apse, instillin an aesthetics of disap rance eye will s e but once.
Delic e and allu ng, a dan rous pleasure ove omes th body rred visi n
further disint rates what is s n (Alaimo 2016, 168). A sm nebula of w e mist is
emi ed fro the process, sl ly float g away, be oming ind guishable from th
sea aro d within sec ds. Watc ng this con ual, smoo ansformation b tween
at is in de and outside, m starts o spiral t o (Alaimo 2016, 163). The
s ral na e of thin s dawns on b dy, which spira with a tre ling br in as it
cl sps onto thro , hol s b dy a oat in its tremors. ver sa ly e sconce in a d n,
ody is art of the diss ve, not a d tan bserve . For it is up ag nst
nthropomorp repr ntatio , iolen atta s and d ial of self ood. All these
processes flow into watery world on currents of carbon and oil, altering this world
and this self. As an octopus considering the spiral of a dissolving shell, mind is
necessarily empathetic, aware of the limitation of ow bei g, o the wa ar s
stre ch o t be on th s s all bo y. ware of th fl w of ings, th ubblin spi l,
the slo pr ess of se s and tho hts dis ving. U il the s al fl s ov r,
eme ng the oth way a und, p t-acid ty. A t ght sti ks and c gs to a urren
f a ne sea and re alls members tre bles wit an ex ctation a kn ledge of its
v rabilit

sticky limbs don't stick

to an 'ontological and epistemological zone that is somehow outside the time and space of the anthropocene'

there is no refuge in thought

in the illusion of impermeability

at the crossroads of body

and place

floats kin

unsafe

uncontained

arm regrows into watery surroundings

which enter body in the same movement

enmeshment of flesh and place

completes itself in this acknowledgment

(Alaimo 2016, 1 and 167-68)

As one figuration of the anthropocene among many others that are possible, the exquisite photographs and videos of dissolving shells may perform cultural work, portraying the shift in alkalinity as a vivid threat to delicate yet essential living creatures. Whereas the predominant sense of the anthropocene subject, en masse, is that of a safely abstracted force, the call to contemplate your shell on acid cultivates a fleshy posthumanist vulnerability that denies the possibility of any living creature existing in a state of separation from its environs. The image of the diminutive creature, with its delicate shell dissolving, provokes an intimacy, a desire to hold and protect, even as we recognize that such beings hover as part of the unfathomable seas. The scene of the dissolve demands an engaged, even fearsome activity of scale shifting from the tiny creature to the vast seas. (Alaimo 2016, 167)

VII.

From violet-black seas to azure skies, creative expressions emerge from all things. A story resides in a hydrogen ion drifting not on a sea but in it, in a fleshy arm regrowing itself out of a small body. Agency, creativity and eloquence emerge in endlessly diverse ways, from 'all of nature's constituents from subatomic particles to the stellar formations' (Oppermann 2019, 452). Eloquence lies in a human word, a language of underwater sound, in bioluminescent signals, in a soft eight-armed body communicating with shape-shifting skin. What does it mean when all these narratives unfold towards each other, overflowing and submerging one another? What happens when a single octopus grips onto an old beer bottle or a shell dissolving on acid, and tells a shred of a human story? Is this that shred, the fragment of meaning floating on currents of acid and carbon, on streams of plastic filling this sea

of this permeable softness

this sea overflows all narratives
if not a thought is given
to a delicate shell dissolving
an amber fish's extinction
an octopus eye dimming
if not a link is made
a recognition of kinship
a shared fleshiness

rising

bodies always permeated with channels

seawater always filled with small bodies

the gaps between things always filled with seawater

this porosity

shift these scales

slowly forging
a posthumanist trans-corporeality (Alaimo 2016, 167)
the necessary condition
of earthly processes

reflect it in words

dwell in the dissolve

fully adrift
intrepidly empathetic (Alaimo 2016, 168)
reach
coil
become
something new and frighteningly immersed
spellbound
coalesced
absorbed
in dangerous pleasures (Alaimo 2016, 168)
and countless saltwater drops

VIII.

I
c
o
i
l
e
d
i
n
t
o
t
h
i
s
s
t
o
r
y

ll
m
a
t
e
r
i
a
l
l
y
e
n
t
e
r
e
d
e
y
e

III
i
m
m
e
r
s
e
d
b
o
d
y
i
n
v
i
o
l
e
t
-
b
l
a
c
k

IV
s
h
o
w
e
d
a
l
a
n
g
u
a
g
e
o
f
l
i
g
h
t

V

shrivelled retracted

VI

resisted the dissolve re rowing

VII
c
o
a
l
e
s
c
e
d
i
n
t
o
t
h
i
s
f
l
e
s
h
y
s
e
l
f

VIII
e
m
e
r
g
e
s
a
n
e
w
f
r
o
m
t
h
e
d
i
s
s
o
l
v
e
d
d
i
s
s
o
l
v
e
s
h
a
p
e
s
a
n
e
w
a
n
i
m

a
l
t
e
n
t
a o f
c t f
l o l
e n e
d k s s
h b
y m
l i

References

Alaimo, Stacy. 2012. States of Suspension: Trans-corporeality at Sea. *Interdisciplinary Studies in Literature and Environment* 19 (3): 476–493.

———. 2013. Violet-black. In *Prismatic Ecology: Ecotheory beyond Green*, ed. Jeffrey Jerome Cohen, 233–251. Minneapolis: University of Minnesota Press.

———. 2016. *Exposed: Environmental Politics and Pleasures in Posthuman Times.* Minnesota: University of Minnesota Press.

Dobrin, Sidney. 2021. *Blue Ecocriticism and the Oceanic Imperative.* London and New York: Routledge.

Evans Ogden, Lesley. 2013. Marine Life on Acid. *BioScience* 63 (5): 322–328.

Oppermann, Serpil. 2019. Storied Seas and Living Metaphors in the Blue Humanities. *Configurations* 27 (4): 443–461.

Scheel, David, Stephanie Chancellor, Martin Hing, Matthew Lawrence, Stefan Linquist, and Peter Godfrey-Smith. 2017. A Second Site Occupied by Octopus Tetricus at High Densities, with Notes on Their Ecology and Behavior. *Marine and Freshwater Behaviour and Physiology* 50 (4): 285–291.

CHAPTER 10

Spectral Species in the (Political) Abyss: From Living Fossils to Presumed Extinctions

Stacy Alaimo

One summer, while swimming at Florida's Emerald Coast, my family became captivated by a steady stream of mullet. The thrill of finding ourselves amid their migration encouraged us to create a plural moniker, "mullay," that befit their elegance. Plunging into Callum Roberts's *The Unnatural History of the Sea*, however, revealed that even this coastal area, known for its abundance of wildlife, was a ghost habitat, with many of the species that had been there/would have been there/should have been there, having been rendered extinct, presumably by the prolonged ravages

Many thanks to the editors, Rachel Murray and Vera Fibisan, for their excellent suggestions!

S. Alaimo (✉)
University of Oregon, Eugene, OR, USA
e-mail: salaimo@uoregon.edu

© The Author(s), under exclusive license to Springer Nature Switzerland AG 2024

V. Fibisan, R. Murray (eds.), *Blue Extinction in Literature, Art, and Culture*, Palgrave Studies in Animals and Literature,
https://doi.org/10.1007/978-3-031-69910-8_10

of settler colonialism and extractive capitalism.[1] The stream of mullets, however delightful in and of themselves, were now an uncanny remainder of all the species that had disappeared. Roberts, recalling his dives in the North Seas in the 1980s, which were unexpectedly empty, saw "the ghosts of the disappeared" and was "saddened by what has been lost" (Roberts 2007, 198). Contemplating blue extinctions while reading Roberts' book at the beach became a practice of speculative grief, imagining the absent species. Such speculations transpire not unlike a filmic dissolve—in which the rather lifeless present and a species-abundant past momentarily share the same scene, giving us a glimpse of what was and what could/should still be.

While nearly any part of the planet could be experienced in this manner (where are the flocks of birds so dense that they darken the skies?)[2] such speculative practices may be more crucial for inspiring deep-sea conservation. Considering violet-black (rather than blue) extinctions entails imagining not only the presence, being, relations, and perspectives of particular species, but somehow visualizing the largest biome on the planet—the deep sea, below 200 meters, which occupies 90% of the ocean environment (Alaimo 2013). The darkness, pressure, temperature, and sheer immensity of this biome have long been considered unfathomable, in terms of metrics, mappings, and imaginaries. The Deep Sea Conservation Coalition website tells us that "Scientists have speculated that as many as 10 million species may inhabit the deep sea" (2023). Despite the biodiversity and volumetric scale of the depths—immensities that overwhelm presumptions of control and containment—popular tropes tagging deep-sea creatures as "alien" sever the largest biome on the planet from the reach of anthropogenic harms, by comparing it to distant planets, or, anachronistically, deeming it the last "great wilderness" for exploration. Such

[1] In terms of the history of the "New World," it is obvious that settler colonialism and capitalist extractivism have caused immense harm to Indigenous peoples and their abundant multispecies worlds. Kyle Powys White provides one example of settler colonialism's affects on landscapes: "As a means of carving out settler homelands from indigenous homelands, waves of settlers harnessed industrial means, from military technologies to large-scale mineral and fossil fuel extraction operations to sweeping, landscape-transforming regimes of commodity agriculture." (Whyte 2017, 208). Such transformation of landscapes—along with the deliberate decimation of species that tribal peoples relied on—resulted in genocidal violence toward tribal peoples as well as the extinction of nonhuman species.

[2] See Mark V. Barrow for accounts of the huge flocks of passenger pigeons in the U.S. in the nineteenth century which created "instant darkness," and a lack of sunlight compared to an "eclipse," (Barrow 2009, 97).

popular formulations, which go hand in hand with capitalist extractivism, neocolonialisms, and the heroic masculinity of the Great White Explorer, dodge the crisis of extinction, as the depths are imagined as both unknown to science and magically untouched by human practices. Yet climate change, ocean acidification, mining, industrial fishing, and pollution throughout the global oceans harm myriad species and ecologies, including those of the depths. Moreover, the abyssal zones are distinctive but not separate from other zones, as they are interconnected in multiple ways for example through the benthic creatures who depend on whale falls and the billions of animals who practice diurnal vertical migration.

The deep seas have long been imagined as a place of strange temporalities which spark speculative forays that dodge evolutionary time and extinction epochs. Western imaginaries of the abyss are often paradoxical. Here we will focus on the book-ended figures of sturdy and enduring "living fossils," on one side, and the fleeting specters of presumably extinct, but never identified species, on the other. This chapter contemplates how these two tropes function, within the current moment of accelerating anthropogenic extinctions and threats to abyssal ecologies. Do spectral taxa of the abyss, as they hail from extraordinarily remote distances and temporalities, insist we reckon with the sheer scale of accelerating anthropogenic extinctions, as well as our own situated responsibilities for species that are anything but companions? How are the tropes of spectral species, with their temporal immensity, articulated with urgent and specific political struggles? This chapter considers these questions by examining two divergent deep-sea tropes, both spectral: enduring "living fossils" and presumed but unknown abyssal extinctions. While the strange temporalities and unfathomable ontologies of these figures (dead fossil or living being? extinct or extant?) may seem otherworldly, both figurations can be mobilized against the impending devastation of deep-sea mining.

Living Fossils and Geological Time

Given the formidable challenges to deep-sea exploration and marine biology in the abyssal zone, stretching from 3000 to 6500 meters, as well as in the hadal trenches, which plunge even deeper, life in the depths—or lack thereof—has long been a matter of speculation. Nineteenth-century naturalist Edward Forbes' "Azoic hypothesis" (1843) asserted that life was not possible in the deep, given the extreme pressure, cold, and darkness (Rozwadowski 2005, 137). However, in 1860 the discovery of a "multitude of unfamiliar creatures attached to the Mediterranean cable raised for repair" from "about 1000 fathoms" attracted the attention of both scientists and the public (Rozwadowski 2005, 15). Despite the palpable evidence of these creatures, it was actually the spectral category of "living fossil" that motivated Wyville Thomson and William Carpenter in the late nineteenth century to undertake deep-sea dredging projects that "pass into the modern era of oceanography as Big Science" (Koslow 2007, 18). Thomson was inspired by G.O. Sars' discovery of a new species of crinoid, "belonging to a group until then believed extinct since the Jurassic" (Koslow 18). Thomson and Carpenter "seized upon what seemed an immense opportunity: to discover a fauna perhaps unchanged since the so-called 'Age of Chalk,' the Miocene fauna fossilized in the chalk beds of Britain. Was there the prospect of encountering the marine invertebrate equivalent of a Jurassic Park fauna still extant in the deep sea?" (Koslow 19).

A century or so later, in 1998, Cindy Lee Van Dover hoped to discover living fossils in the deep-sea vents: "It is to these remote ridge systems that I look for vindication of what I think of as the 'trilobite factor'—the belief that there are still major discoveries waiting for us in unexplored regions of the ocean. Trilobites, of course, have been extinct since the Permian, some 250 million years ago, and the chances of finding some deep-water relative of these fossil arthropods teeming about an abyssal hot spring are perhaps remote to nil. But, if not trilobites, surely other unimagined celebrations of nature's ingenuity will be found" (Van Dover 1998, 438). Wild hopes for discovering living fossils morph into a more capacious desire to find other marvels. The enduring desire for the deep sea to be an exceptional domain where living fossils lurk and linger is notable. Although the concept of "living fossil" is debated in science and philosophy, the figuration's specific mobilizations and broad appeal may impact whether deep-sea species and biomes register as "worthy" of concern. When deep-sea creatures circulate in contemporary popular culture as "living fossils," they suggest a conflation of time and space, as they are flung back into earlier geological epochs, doubly distant from human domains. Such imaginaries may serve to reassure us that there are ecosystems and species that remain safe from human harm, which is not exactly helpful for ocean conservation.

When Charles Darwin refers to living fossils, he does so cautiously: "These anomalous forms may almost be called living fossils; they have endured to the present day, from having inhabited a confined area, and from having thus been exposed to less severe competition" (Darwin 1859, 107). Note: "may almost be called." Evoked without such delicacy, the oxymoron, notoriously persisting without definition and promiscuously circulating in sensationalized forms, invites outright rejection as illogical and unscientific. Didier Casane and Patrick Larenti, for example, baldly conclude, "We hope that this review will contribute to dispelling the myth of the coelacanth as a 'living fossil' and help biologists keep in mind that actual fossils are dead" (Casane and Larenti 2013, 337). Similarly, Mark Carnall, Zoological Collections Manager, exhorts, "Let's Make Living Fossils Extinct," charging that the murky concept causes misunderstanding (Carnall 2016). Other scientists and philosophers, however, defend the concept.

Paleontologist and marine ecologist Scott Lidgard and philosopher of science Alan C. Love, in "Rethinking Living Fossils," reanimate the ostensibly dead concept, contending that it provokes novel theoretical research:

> Biologists would be mistaken if they relegated living fossils to paleontological inquiry or assumed that the concept is dead. … Instead of viewing the concept's task as categorizing living fossils, we show how its primary role is to mark out what is in need of explanation, accounting for the persistence of both molecular and morphological traits. Rethinking different conceptions of living fossils as specific hypotheses reveals novel avenues for research that integrate phylogenetics, ecological and evolutionary modeling, and evo-devo to produce a more unified theoretical outlook. (Lidgard and Love 2018, 760)

In this defense, the concept underscores "prolonged stasis" as a question requiring investigation from multiple disciplinary and methodological perspectives. Interestingly, even the crudest accounts of "living fossils" in popular media, as we will discuss below, sensationalize the "weirdness" of this prolonged stasis, inviting the public to participate into the bafflement, if not the research. In terms of abyssal life, the twenty-first century acceleration of the discovery of "new" species or taxa that persist across vast time scales plays out—hauntingly—against the accelerating threats to the now precarious life in the deep. Weirdness can provoke interest, but exactly how the concept of living fossils is articulated and mobilized in popular media, as we will discuss below, matters.

Philosopher Derek D. Turner defends the category of living fossils for science and conservation, noting its "practical and theoretical advantages."

Living fossils, Turner argues, have three sometimes co-occurring features: "Prehistorically deep morphological stability; Few extant species; and High contribution to phylogenetic diversity" (Turner 2019, 3). Turner then opens the question to conservation, "One might think that the loss of a coelacanth or horseshoe crab species would be especially tragic, that it would be worse than other extinctions. My claim is that the phylogenetic living fossil concept helps explain why" (3). Not only should the living fossil concept "guide empirical research," it should play a role in "setting conservation priorities." In other words, "we should appropriate the living fossil concept, putting it to work in a way that reflects going concerns about biodiversity loss" (3, 5). Moreover, he explicitly argues for the value of creatures such as the horseshoe crab that create "the sense of being placed into contact with the deep past" (10). As we analyze how abyssal living fossils circulate in popular media, it is important to note that even as the term is often misused, it has both scientific validity and cultural impact: it marks the importance of particular taxa, species, and genes, and it has the potential to cultivate connections to deep time. For abyssal life, the trope of the living fossil can connect us not only to deep time but with unfathomably deep places, explicitly or implicitly arguing that they too are within the terrain of environmental concern. Deep-sea living fossils, with their paradoxical endurance and precarity, dwell across times and places that are unthinkably distant from contemporary terrestrial terrains. They are simultaneously symptoms of and remedies for the maladies attendant upon environmental subjects grappling with the now familiar problems of what has been termed the Anthropocene—problems of scale, culpability, agency, extinction, and more.

Claire Nouvian's magnificent coffee table book, *The Deep: Extraordinary Creatures of the Abyss* (2007), which features photographs of abyssal life and accessible scientific essays, designs aesthetic and affective lures to concern. One of these essays, by evolutionary biologist Robert C. Vrijenhoek, entitled "Are Deep-Sea Animals Living Fossils?" notes the discovery of the coelacanth in 1938, which magnified the wish that the deep sea could be harboring more living fossils. Alas, he notes, during the 1970s scientists "hoped that they might find 'living fossils' at vents, perhaps even a living trilobite, a group of arthropods that dominated the Paleozoic seas [543–248 million years ago]," but they were disappointed (Vrijenhoek 2007, 128). Vrijenhoek notes that although few "true living fossils have been recognized out of the thousands of deep-sea species identified during the past century," one species epitomizes the category—the vampire squid: "it has changed relatively little in hundreds of millions of years." He credits its "ability to live in oxygen-poor waters" with its survival for "millions of years while so many other deep-sea animals have gone extinct" (127). Another essay in *The Deep*, "The Vampire Squid from Hell," explains why it is exceptional: "*Vampyroteuthis infernalis* is one of the rare, authentic, living fossils that find shelter in the deep sea; its origins date back over 200 million years. It may even be the common ancestor of octopuses and squids because it displays characteristics of both groups." This "archaic creature" can "reside permanently within the oxygen minimum layer," due to a "respiratory blood pigment." It also has the slowest "metabolism of any cephalopod," which "allows it to go about its routine in an environment that is utterly hostile to other species" (Nouvian 2007, 134–135). The vampire squid hails from an incredibly distant time and place, and yet Nouvian positions the photo of the vibrant orange squid close-up, with its big blue eye aimed at the person holding the book, as the metonymic abyss becomes surreally at hand (Alaimo 2025). Kantaro Fujioka and Dhugal Lindsay in "Deep Trenches: The Ultimate Abysses," included in *The Deep*, express their astonishment at finding foraminiferans in the trenches: "Foraminiferans are unicellular animals and an important link in the food chain in many environments. Finding abundant colonies of these living fossils alive at the deepest point on Earth was overwhelming" (Fujioka and Lindsay 2007, 241). The epochal endurance of living fossils intensifies scientists' sense of wonder, even astonishment, which spills over to popular audiences.

Popular digital magazines and other web sites also feature "living fossils" from the deep, usually as click bait. Ironically, the depiction of the "alien abyss" is all too human, as the photos of sensationalized life forms are buried in a barrage of advertisements—popping up all over—as capitalism invades the farthest reaches of the planet, crowding out creaturely life with its rude, distracting, energy. "'Living Fossil' Thought Extinct for 273 Years Found Thriving on Ocean Floor," by Michelle Star, in *Science Alert*, begins with a photo of a strange specimen, looking like an orange anemone attached to a snake. These "benthic besties" turn out to be a crinoid and a coral in a symbiotic relationship (Star 2021). The article "'Living Fossil' Deep Sea Shark Pregnant with Six Pups Pulled from Ocean," by Jess Thomson, in *Newsweek*, tells a grotesque tale, in which a modicum of maternalist sympathy is invoked for this rare animal "captured" before she "gave birth to her six young." Then she is demonized, as a "monster of the deep," with a "characteristically ugly face," and "evil fangs." Finally, and bizarrely, the story concludes with a racist environmentalist message, explaining that this "rare prehistoric deep-sea shark," a "'Goblin shark,'" was "caught accidently" by "bottom trawlers" near Taiwan, where the practice is legal. Somehow, after the sloppy references to the Goblin shark as being prehistoric and the invocation of the ugly and monstrous, the story concludes as a preachy environmental cautionary tale: "This fishing practice is banned in many regions of the oceans due to the effects it has on seafloor species and the amount of bycatch it causes" (Thomson 2023).

With better graphics and fewer frenzied ads, the story "Living Fossil Discovered During Deep Sea Expedition Off of Queensland Coast," by Mark Rigby and Adam Stephen, underscores the researchers' astonishment: "they had no idea they would discover living fossils, ecosystems that have remained unchanged for millions of years and marine life thought alien to Australian waters" (Rigby and Stephen 2016). Oddly, the story refers to the ecosystems as "living fossils," even though one of the scientists distinguishes creatures from their habitats, saying that the sea lilies they found were "relict animals that have existed through millions of years because they're in this nice stable environment." And while the story contains photos of other species they found, it doesn't clarify which of these species (or taxa) qualify as "living fossils," noting instead, the surprise of finding species from other regions in Australian waters: coral reef fish (Tahiti), the sea toad (South America), and precious coral (the Mediterranean). The scrambling of habitat and taxa as well as the jumbling of time and space suggest the sticky appeal of the concept of the deep-sea living fossil—the trope itself endures, surfacing even where it makes little sense. Given the early definitions of "abyss" as primordial chaos, the confusions and conflations within this article perform as enticements to a mythical sense of mysterious seas, drowning the environmental anxieties of the current moment by plunging us into an abundant, marvelous, but murky realm.

While the stasis of "living fossils" has been seized upon by creationist groups as "evidence" against evolution, their endurance has also been invoked to call attention to accelerated anthropogenic extinctions. Evolutionary biologist Alice Clement, in "From Coelacanths to Crinoids: These living fossils haven't Changed for 9 Million Years," features dramatic images and descriptions, explaining that "their slowing rates of evolution are a likely result of the relatively stable environment" of the deep sea (Clement 2022). The refuted "azoic" theory lives on, concretized into animated fossils, a sort of living dead. The striking photos in the essay alternate between (dead) fossils on display and living creatures, many of whom are photographed in their natural habitats. The caption of the sliced nautilus specimen, exhibiting its spiraling inner chambers, notes, "You'd struggle to tell an ancient Nautilus from a living one." For those of us grieving extinctions, the endurance of these "living fossils" may be comforting, even hopeful. Clement concludes, however, with a vague and impotent warning: even these species are "not immune," and if "we're not careful, we may lose some of these creatures forever" (Clement 2022). The call turns on the paradoxical conflation of endurance and endangerment, resilient stasis and rapidly accelerating threats.

Other stories employ living fossils as environmental cautionary tales, object lessons saturated with and amplified by deep time. Dramatically, Alexander Haro's article, "A Living Fossil From the Deep Sea Was Discovered With Trash Crammed In its Stomach," includes a vivid photograph of a legendary living fossil, the coelacanth, with the bright green edges of a Lay's potato chip bag crinkling out of its gut. The essay begins by explaining that coelacanths were believed to be extinct until Marjorie Cortenay Latimer discovered one in the bycatch of South African fishermen. "Turns out they were just hiding in the deepest bits of the ocean remaining virtually unchanged for 400 million years. Now, they're eating potato chip bags" (Haro 2018). Sophie McDonald's essay, "Nautilus: The Living Fossil Under Threat in Our Modern World," encourages readers to imagine an "underwater Jurassic Park ... filled with creatures of all shapes and sizes from reptilian beasts to burrowing worms." A gorgeous photo of a chambered nautilus follows, the caption explaining that it has "remained unchanged for millions of years" (McDonald 2020). The aesthetic appeal of the nautilus, however, may threaten their existence, as their shells are sold as decorations. The living fossil underscores the enormity of the current wave of anthropogenic extinctions: "The chambered nautilus existed before dinosaurs and has managed to survive every major global extinction event, only to be decimated by humans in the Anthropocene. This bizarre and beautiful species deserves more than a future on someone's mantlepiece, nautili should continue to inhabit our oceans for eons to come" (McDonald). Living species designated as "fossils" perform as potent bids for imagining abundant futures. Even the deep-sea "ghost shark," which is not considered endangered, receives attention as a living fossil deserving protection from future threats. Save Our Seas Foundation proclaims: "Amazingly, the Australian ghost shark has been shown to have the slowest evolving genome of all known vertebrates and is thus a 'living fossil' (even more so than the coelacanth!), giving scientists great insights into the evolutionary history of fish and vertebrates." Focusing on how noise pollution may be affecting the ghost shark's ability to hear, Save Our Seas has launched "The Ghost of Oceans Future," cleverly condensing the hope that the ghost shark will survive into the future with the spectrality of the warning that it may not (Save Our Seas 2023). As McCorristine and Adams argue, "spectral ideas ... shape the way conservationists think and act: in response to past and present extinctions, money is spent, organizations and projects are created, and land is surveyed and set aside for 'nature.' People are recruited to conservation causes by stories of past extinctions and about those predicted to occur in the future" (McCorristine and Adams 2020, 104). Spectral tropes of deep-sea life are especially potent for shaping marine conservation, as they plunge publics into an abyss of speculation not only about particular species, but about confluences between the limits of human knowledge and the immensity of anthropogenic harms.

Unknown Extinctions in the Abyss

Living fossils somehow endure through geological epochs and extinction events. They circulate as both a balm to concerns about anthropogenic extinctions, and, conversely, as a call to protect these exceptional taxa by decelerating the global destruction of marine life. Metonymically, the solidity of fossils contradicts the precarity of the living beings.[3] The coelacanth displayed in museums and the nautilus decorating a mantlepiece are tangible entities captured by science and consumerism, physically present, valued as objects if not as beings. They place us in contact with deep time, as Turner argues, seemingly dissolving the distance between ordinary domestic habitats and the unfathomable life worlds of these creatures. Spectral species on the other hand are those who are merely presumed to have existed, who have been rendered extinct before being discovered, and who have not been captured in photos, videos, or as physical specimens. They seem to have left no trace, and yet, still, they must have existed. What sorts of cultural work can the specters of presumed

[3] The seeming solidity of living fossils, especially those with shells, contrasts with the effects of ocean acidification on the shells of many species including pteropods, which are vital for ocean ecologies. See my chapter, "Your Shell on Acid: Material Immersion, Anthropocene Dissolves," in *Exposed* (Alaimo 2016).

extinction—the species conjured up by reasonable speculations, the ones that must have existed, must have been rendered extinct—accomplish? Geoffrey Bowker writes that "things that do not get classified are not considered of economic, aesthetic, or philosophical importance," and yet they may exist as part of the "shadow systems to the exclusions" (Bowker 2005, 153). The specters of presumed abyssal extinctions may dwell in shadow systems, but in this case, however, those systems would have been their own watery life worlds, not the structures of human knowledge. Moreover, these spectral species, however wispy, may actually be of immense aesthetic and philosophical importance, as they spark ethical and political visions, commitments, and initiatives that would hold us accountable even for the extinctions not yet recorded, nor ever to have been recorded.

The amorphous specter of unknown extinctions flows through various genres, disciplines, and media pertaining to the abyss. Such apparitions often float hand in hand with conservationism. Roberts writes in the chapter, "Violating the Last Great Wilderness": "For decades now deep-water trawlers have roamed the high seas, destroying habitats of incalculable biological significance; places that would be treasured in national parks if they were on land." He concludes: "We are losing life in the deep sea before it has even been described by science" (Roberts 2007, 29). Even a prosaic McGraw Hill textbook, *Marine Biology*, speculates about the extinction of undiscovered species: "With some 30,000 seamounts in the deep sea, nearly all unexplored, there must be vast numbers of seamount species waiting to be discovered, but may go extinct before that happens" (Castro and Huber 2008, 375). The conceit that creatures are "waiting to be discovered" suggests they have read Bruno Latour and are awaiting their representation within a new *Politics of Nature (2004)*, or, less academically, that they are pining for their 15 minutes of fame (Latour, 2004). Yet, the technologies of abyssal "discovery," which include noise and bright lights, are disruptive, if not violent and deadly in and of themselves. Nonetheless, the far-reaching damages of the Anthropocene require both scientific captures and their broad dissemination in order to provoke global public concern and the will to protect abyssal, pelagic, and other marine life.

The decade-long global science project, The Census of Marine Life, undertook the impossible task of counting and identifying all the species in the sea during the first decade of the twenty-first century. Despite the project's laudable ambition, Census scientists realized that species would elude their grasp. Paul V.R. Snelgrove, in a section on "The Unknowable," from the small coffee table book, *Discoveries of the Census of Marine Life: Making Ocean Life Count*, writes: "We will eventually know the identities of most species in most environments, the distribution of many species in many environments, and abundances of common species. But the size of the ocean, the remoteness of the deep-sea biome, and the limited global scientific capacity will doom some species to remain unknown, perhaps forever. Some of these species will go extinct without us ever seeing them and others will continue to quietly do whatever it is they do in Planet Ocean" (Snelgrove 2010, 241). Snelgrove dodges the question of *why* these species will go extinct, making it a *fait accompli,* a future that cannot be prevented. Snelgrove's placid, desiccated stance of cold objectivity is not adequate for the ethical and political quandaries in this moment of anthropogenic extinctions. As Astrida Neimanis eloquently argues, "we must find ways to call ourselves into account, to enact an ethics of curiosity and care, to do politics even if we know they are always incomplete." She adds that this "might in fact be *the* challenge of the Anthropocene" (Neimanis 2020, 59).

Suddenly Disappearing into Debris: Deep-Sea Extinctions in Political Time

The looming catastrophe of deep-sea mining demands that we call ourselves into account, practice an ethics of curiosity and care, and engage in politics, however incomplete, as Neimanis argues. The rapid decimation of immense benthic habitats and an astonishing number of species, as well as likely harms to wider marine ecosystems, combined with incomplete scientific understanding of the ocean floor, pose deep-sea mining as a quintessential quandary of the Anthropocene, as well as a "wicked" environmental problem (Rittel and Webber 1973), given that some of these minerals could be used to make batteries for electric cars to combat climate change. Michelle Bastian, writing of the effect of whaling on benthic life, argues that even the unknown extinctions can be storied, by rewriting "their loss as a situated one, a situated, unknown extinction, grounded within specific processes of habitat loss and resource extraction, with specific stories of benefit, loss, and responsibility that we still must unravel" (Bastian 2020, 469). While my work has long been informed by Donna Haraway's feminist epistemology of "situated knowledges," which entail "partial perspectives" and responsibility (Haraway 1988), I don't think it is possible that the vast majority of undiscovered species who have been rendered extinct by human activities can be storied in this manner; nor do I believe that narratives are the most potent force for eliciting concern for marine life. My book, *The Abyss Stares Back,* argues that aesthetic encounters with deep-sea life—beings that elude anthropomorphic narratives—can spark environmental concern. Situated, trans-corporeal tracings of benefit, loss, and responsibility become more feasible, however, within pivotal political moments, such as the current moment of looming deep-sea mining. One can imagine how benthic and midwater marine life will be suddenly destroyed by the giant nodule "collection" machines that will damage the seafloor, fill the area with moving "clouds" of matter, and instantly or slowly transform living creatures into mere spectrality. Sue Reid describes the vulnerabilities of benthic life: "Moving up through the water column to escape mining plumes is no option for slow-growing, slow-moving deep-sea creatures whose only escape is sideward across the ocean floor and bottom layer of seawater" (Reid 2020, 40).

Disturbingly, a two-minute animated video produced by MIT engineering depicts myriad benthic creatures as having already disappeared. Richard Grusin explains that the "logic of premediation insists that the future itself is already mediated and with the right technologies … the future can be remediated before it happens" (Grusin 2004, 19). The *mise en scene* of this animated video premediates the disappearance of all living creatures on the seafloor and in the midwater above, rendering vast areas as inert "natural resources" for exploitation. Included in Alexandra Gillespie's story for NPR, "Your Next Car May be Built with Ocean Rocks. Scientists Can't Decide if that is Good" (Gillespie, 2021), the animation depicts a ship lowering a "collector vehicle," descending 4000 meters to the sea floor, the process of collection, and the resulting debris plumes. The voiceover describes a plume consisting of "unwanted, unneeded" material that would be "returned to the ocean," either "in the midwater column or back down in the deep ocean" (MIT 2019). How many living beings would be part of that "unwanted" "material"? The detached voice ignores the dire consequences—to particular living beings—that the choice between midwater or seafloor would entail. Most disturbingly, however, even though a cool green submarine cruises by the mining scene, for no apparent reason, absolutely no ocean life is included in this video. The seafloor—the scene of the mining—is represented as already utterly devoid of life. Benthic species are rendered nonexistent in this ghastly "animation."

By contrast, a *New York Times* multimedia story, "Deep Sea Riches: Mining a Remote Ecosystem," which includes the section, "Life Among the Nodules," explains: "Polymetallic nodules are an anchor for a fragile and slow growing ecosystem that includes species found nowhere else on earth" (Imbler and Corum 2022). A video zooms in on "several glass sponges" growing on a nodule; another video shows an "unidentified species of jellyfish" hovering right above the seafloor, followed by videos and photos of fascinating species, on or just above the seafloor, including the "ping-pong tree sponge," and a glittering, fluttering squidworm. Having been captivated by these gorgeous creatures, it is disturbing to find oneself watching a subsequent video of a rapidly expanding sediment cloud—which positions the viewer within the powerful wave of disrupted sediment. This visualization dramatizes the powerful force of the sediment plumes, whereas the MIT animation depicted the plumes as benign and contained. Another potent digital work is the data visualization, "Deep-Sea Mining May Harm Thousands of Species Before They Are Even Discovered," which depicts a map of the Clarion Clipper Zone, with overlays of the "exploration areas," and "areas of particular environmental interest," noting that new research suggests that the APEI's "may not protect the species most likely to be affected" by mining (Pew 2023). The geometrical overlay of mismatches is concerning. But they are followed by a delightful surprise—cartoonish and oversized drawings of various benthic creatures—seemingly placed on top of the map. Their vibrant colors and intriguing shapes, not to mention their comic oversizing, implicitly perform a recognition of the value of living, animate(d) beings—a protest against the deadening feats of large-scale extractivism.[4]

Lara Macheriotou et al., in a scientific article about the potential effects of deep-sea mining in the Clarion-Clipperton Fracture Zone, conclude not only that the "taxa of these nodule-bearing abyssal plains will be especially vulnerable to the risks of extinction," but that "sediment plumes released into the water column of the CCFZ from mining activities are expected to persist for many years," affecting both the benthic and pelagic realms (Macheriotou 2020, 1, 8). Karen McVeigh, in *The Guardian*, sounds a similar alert: "Deep Sea Mining May Push Hundreds of Species to Extinction, Researchers Warn" (McVeigh 2021). The vulnerability of

[4] The large fish added to the map may also be a wry reference to maps such as the medieval Carta Marina (1539), which populates the seas with gigantic monsters. Thanks to Rachel Murray and Vera Fibisan for this insight.

marine life to anthropogenic extinction saturates otherwise sanguine accounts of scientific discovery with sadness. Dr. Diva Amon, a deep-sea biologist, in the guest post "Megafauna and Minerals on the Pacific Abyss," published in *Deep Sea News*, tells of seeing pretty "AH-MAZING" sights, such as "an anemone-like animal with 8-foot tentacles that billow across the seafloor." Amon explains that this creature, "*Relicanthus* sp., is so different from other anemones it was recently moved to a new order" (Amon 2016). A lovely photo of the *Relicanthus* gliding across the abyssal plain follows. After pausing to appreciate the delicate beauty of this remarkable creature, with its long, wispy tentacles barely visible, we read on: "As incredible as seeing this tentacled beast was, I couldn't help but feel a tinge of sadness. It's difficult for a marine biologist working in an area that may be forever changed within the next two decades. As the demand for metals increases, humans are seeking resources in ever more remote places and the next frontier of mining will likely take place in the deep ocean" (Amon). Readers may empathize with the marine biologist as well as with the creature. The next photo shows two human hands holding the polymetallic nodules, presenting a dramatic contrast. The newly discovered anemone seems already spectral in its fragility, whereas the solid nodules, held in the hands that routinely hold cell phones and other technology for which the minerals are mined, are invulnerable objects. One reader comments: "How can we, as citizens, register our enormous and overwhelming distaste for this sort of exploitation? This is distressing to an extreme degree."

Political agency is a maddening problem for environmentalists as well as for theorists. Even as we qualify the whitewashing species designation of the "anthro" of the Anthropocene to stress capitalism, colonialisms, racism, and extractivism, the question of the massive scale of "human" harms to planetary systems, nonhuman species, as well as particular groups of people remains. How do "we" begin to make sense of the destruction of abyssal zones that are unimaginably distant and different from terrestrial life and yet are being/will be decimated by such things as the banal consumerism of buying electronics or eating orange roughy for dinner? Deep-sea mining could begin in 2025. By the time you read this essay—unless the political pressure is so strong as to reverse course, it may have already begun. Scientists, scholars, writers, ocean conservation groups, Indigenous peoples, and popular media have sounded their warnings and expressed their opposition.

Spectral species of the abyss, as oxymorons, mind-warping temporalities, speculative unknowability, and strangely fragile solidity, call us to navigate the turbulent places where multiple temporalities affect ethical and political agency. Sue Reid's potent essay, "Solwara 1 and the Sessile Ones," concludes by stressing the need for ocean legalities to attend to the distinctive temporalities of abyssal species and their habitats. "There is a compelling place in law to better recognize the different registers of temporality. Legal apparatuses must be put in place that acknowledge life stages and associated vulnerabilities of creatures on the same plane as the commercial claims of corporations" (Reid 2020, 41). I agree with Reid's call for legal systems attuned to sustaining living beings and their habitats rather than corporations. Another vital temporal framework comes from Indigenous cultures, especially, in this case, those in the Pacific who have signed onto "Indigenous Voices for a Ban on Deep Sea Mining," which states, "For millennia our people have lived in a relationship with the natural world that is defined by respect, gratitude, responsibility, and love. Our genealogies, woven across space and time, connect us physically and spiritually to animals and plants from the highest mountains to the deepest oceans" (Indigenous Voices 2023). Cultures that have practiced intimate and respectful relations with the natural world—for millennia—should be the ones the rest of the world listens to as the killing machines of rapid, greedy, extractive capitalism rev up their engines.

Shane McCorristine and William Adams, in "Ghost Species: Spectral Geographies of Biodiversity Conservation," ask whether "being 'haunted' by species accomplish[es] anything?" (2020, 101). I ask in response, can we invite spectral species to haunt us in ways that will expand ethical and political terrains of concern? Can we embrace the strange temporalities of spectral species of the abyss, without resignation or despair, but instead, as an incitement to intensify our sense of political time even as it is embedded in vaster time scales of evolution and anthropogenic harm? A vibrant, ecologically functioning planet must hold more than fossils.

References

Alaimo, Stacy. 2013. Violet-Black: Ecologies of the Abyssal Zone. In *Prismatic Ecologies: Ecotheory Beyond Green*, ed. Jeffrey Jerome Cohen, 233–225. Minneapolis: University of Minnesota Press.

———. 2016. *Exposed: Environmental Politics and Pleasures in Posthuman Times.* Minneapolis: University of Minnesota Press.

———. 2025. *The Abyss Stares Back*. University of Minnesota Press. forthcoming.

Amon, Diva. 2016. Megafauna and Minerals on the Pacific Abyss. *Deep Sea News*.. https://deepseanews.com/2016/01/megafauna-and-minerals-on-the-pacific-abyss/. Accessed 5 April 2024.

Barrow, Mark V. 2009. *Nature's Ghosts: Confronting Extinction from the Age of Jefferson to the Age of Ecology*. Chicago: University of Chicago Press.

Bastian, Michelle. 2020. Whale Falls, Suspended Ground, and Extinctions Never Known. *Environmental Humanities* 12 (2): 454–474.

Bowker, Geoffrey C. 2005. *Memory Practices in the Sciences*. Cambridge: MIT Press.

Casane Didier, and Laurenti Patrick. 2013. Why Coelacanths are not 'Living Fossils': A Review of Molecular and Morphological Data. *Bioessays*. 35 (4): 332–38.

Carnall, Mark. 2016. Let's Make Living Fossils Extinct. *The Guardian*.. https://www.theguardian.com/science/2016/jul/06/why-its-time-to-make-living-fossils-extinct. Accessed 5 April 2024.

Castro, Peter, and Michael E. Huber. 2008. *Marine Biology*. 7th ed. Boston: McGraw-Hill.

Clement, Alice. 2022. From Coelacanths to Crinoids: These Living Fossils Haven't Changed for 9 Million Years. *The Conversation*. https://theconversation.com/from-coelacanths-to-crinoids-these-9-livingfossils-havent-changed-in-millions-of-years-188886. Accessed 5 April 2024.

Darwin, Charles Robert. 1859. *A Facsimile of the First Edition of On the Origin of Species*. Cambridge: Belknap Press.

Deep Sea Conservation Coalition. 2023. https://savethehighseas.org/discover-thedeepsea/#:~:text=The%20deep%20sea%2C%20often%20defined,largest%20biom e%20on%20the%20planet. Accessed 5 April 2024.

Fujioka, Kantaro, and Lindsay, Dhugal. 2007. Deep Trenches: The Ultimate Abysses. In *The Deep: The Extraordinary Creatures of the Abyss*, 240–41. Chicago: University of Chicago Press.

Gillespie, Alexandra. 2021. Your Next Car May be Built with Ocean Rocks. Scientists Can't Decide if that is Good. *NPR*. September 3. https://www.npr.org/2021/09/03/1031434711/your-next-car-may-be-builtwith-ocean-rocks-scientists-cant-agree-if-thats-good. Accessed 5 April 2024.

Grusin, Richard. 2004. Premediation. *Criticism* 46 (1): 17–39.

Haraway, Donna J. 1988. Situated Knowledges: The Science Question in Feminism and the Privilege of Partial Perspective. *Feminist Studies* 14 (3): 575–599.

Haro, Alexander. 2018. A Living Fossil from the Deep Sea Was Discovered with Trash Crammed in its Stomach. *The Inertia*. https://www.theinertia.com/environment/coelacanth-lays-chip-bag-in-stomach/. Accessed 5 April 2024.

Imbler, Sabrina, and Corum, Jonathan. 2022. Deep Sea Riches: Mining a Remote Ecosystem. Multimedia. *New York Times*.. https://www.nytimes.com/interactive/2022/08/29/world/deep-sea-riches-mining-nodules.html. Accessed 5 April 2024.

Indigenous Voices for a Ban on Deep Sea Mining. n.d.. https://www.blueclimateinitiative.org/say-no-to-deep-sea-mining. Accessed November 2023.

Koslow, Tony. 2007. *The Silent Deep: The Discovery, Ecology, and Conservation of the Deep Sea*. Sydney: UNSW Press.

Latour, Bruno. 2004. *Politics of Nature: How to Bring the Sciences into Democracy*. Trans. C. Porter. Cambridge: Harvard UP.

Lidgard, Scott, and Alan Love. 2018. Rethinking Living Fossils. *Bioscience* 68 (10): 760–770.

Macheriotou, Lara, et al. 2020. *Phylogenetic Clustering and Rarity Imply Risk of Local Species Extinction in the Prospective Deep-Sea Mining Areas of the Clarion Clipperton Fracture Zone*, 1–10. Royal Society Publishing Proceedings B.

McCorristine, Shane, and William M. Adams. 2020. Ghost Species: Spectral Geographies of Biodiversity Conservation. *Cultural Geographies* 27 (1): 101–115.

McDonald, Sophie. 2020. Nautilus: The Living Fossil Under Threat in Our Modern World. https://www.themarinediaries.com/tmd-blog/nautilus-the-living-fossil-under-threat-in-our-modern-world. Accessed 5 April 2024.

McVeigh, Karen. 2021. Deep Sea Mining May Push Hundreds of Species to Extinction, Researchers Warn. *The Guardian*.. https://www.theguardian.com/environment/2021/dec/09/mining-may-pushhundreds-of-deep-sea-mollusc-species-to-extinction. Accessed 5 April 2024.

Star, Michelle. 2021. "Living Fossil" Thought Extinct for 273 Years Found Thriving on Ocean Floor. *Science Alert*.. https://www.sciencealert.com/living-fossilthought-extinct-for-273-million-years-found-thriving-on-the-ocean-floor. Accessed 5 April 2024.

Neimanis, Astrida. 2020. Held in Suspense Mustard Gas Legalities in the Gotland Deep. In *Blue Legalities: The Life and Laws of the Sea*, ed. Irus Braverman and Elizabeth R. Johnson, 45–62. Durham: Duke University Press.

Nouvian, Claire. 2007. *The Deep: The Extraordinary Creatures of the Abyss*. Chicago: University of Chicago Press.

Pew Trust. 2023. Deep Sea Mining May Harm Thousands of Species Before They Are Even Discovered. Data Visualization. https://www.pewtrusts.org/en/research-and-analysis/datavisualizations/2023/deep-sea-mining-may-harm-thousands-of-species-before-they-are-even-discovered#:~:text=Deep%2DSea%20Mining%20May%20Harm,Discovered%20 %7C%20 The%20Pew%20Charitable%20Trusts. Accessed 5 April 2024.

Reid, Susan. 2020. Solwara 1 and the Sessile Ones. In *Blue Legalities: The Life and Laws of the Sea*, ed. Irus Braverman and Elizabeth R. Johnson, 25–44. Durham: Duke University Press.

Rigby, Mark, and Adam Stephen. 2016. Living Fossil Discovered During Deep Sea Expedition Off of Queensland Coast. *ABC North*. https://www.abc.net.au/news/2016-01-20/living-fossils-discovered-during-deep-sea-expedition-off-qld/7101224. Accessed 5 April 2024.

Rittel, Horst W., and Melvin M. Webber. 1973. Dilemmas in a General Theory of Planning. *Policy Sciences* 4 (2): 155–169.

Roberts, Callum. 2007. *The Unnatural History of the Sea*. Washington: Island Press.

Rozwadowski, Helen M. 2005. *Fathoming the Ocean: The Discovery and Exploration of the Deep Sea*. Cambridge: Belknap Press.

Snelgrove, Paul. 2010. *Discoveries of the Census of Marine Life: Making Ocean Life Count*. Cambridge: Cambridge UP.

The Ghosts of Oceans Future. Save Our Seas Foundation. 2023. https://saveourseas.com/project/the-ghost-ofoceans-future/. Accessed 5 April 2024.

Thomson, Jess. 2023. 'Living Fossil' Deep Sea Shark Pregnant with Six Pups Pulled from Ocean. *Newsweek*.. https://www.newsweek.com/rare-goblin-shark-found-babies-stomach-1807288. Accessed 5 April 2024.

Turner, Derek D. 2019. In Defense of Living Fossils. *Biology and Philosophy* 34 (23): 1–22.

Van Dover, Cindy Lee. 1998. Vents at Higher Frequency. *Nature* 395: 437–438.

Vrijenhoek, R.C. 2007. Are Deep Sea Animals Living Fossils? In *The Deep: The Extraordinary Creatures of the Abyss*. Chicago: University of Chicago Press.

Whyte, Kyle Powys. 2017. Indigenous Conservation and the Anthropocene. In *The Routledge Companion to the Environmental Humanities*, ed. Ursula Heise, Jon Christensen, and Michelle Niemann, 206–215. New York: Routledge.

CHAPTER 11

Alien Rhythms: Sounding Black Futures from the Ocean Floor

Henry Ivry and Max Karpinski

> The ships landed long ago: they already laid waste whole societies, abducted and genetically altered swathes of citizenry, imposed without surcease their values. Africa and America—and so by extension Europe and Asia—are already in their various ways Alien Nation. No return to normal is possible: what 'normal' is there to return to? Part of the story of black music (the affirmative, soul-gospel aspect) has always been this—that losing everything except basic dignity and decency is potentially a survivable disaster. The other part—as told so obliquely by Ra, Coltrane, Braxton, Delany, Ishmael Reed, and doubtless many others less easily seen than this—is that staying true to the best in yourself may mean when everything can so cunningly imitate everything else, talking in dark, crazed, visionary tongues for a season. (Sinker 1992, 33)

H. Ivry (✉)
University of Glasgow, Glasgow, UK
e-mail: henry.ivry@glasgow.ac.uk

M. Karpinski
York University, Toronto, ON, Canada
e-mail: maxkarp@yorku.ca

© The Author(s), under exclusive license to Springer Nature Switzerland AG 2024

V. Fibisan, R. Murray (eds.), *Blue Extinction in Literature, Art, and Culture*, Palgrave Studies in Animals and Literature,
https://doi.org/10.1007/978-3-031-69910-8_11

The Water Line

Early in his recent collection of essays that thinks through the relations between life, water, and sound—a key triad for the chapter that follows—Stefan Helmreich articulates a provocative claim using a somewhat startling turn of phrase. Parroting the language of W.E.B. Du Bois's 1903 *The Souls of Black Folk*, where he argues "the problem of the twentieth century is the problem of the color-line" (qtd. in Helmreich 2016, xxi), Helmreich asserts the urgency and necessity of the blue humanities in the contemporary moment. "The problem of the twenty-first century," he writes, "is the problem of the water line" (xxi). We might describe this echo as an instance of *bluewashing*, a moment in which Du Bois's language and critical analytic are both ported over, recontextualized, and deemed equally applicable to Helmreich's three guiding concepts—life, water, and sound—without any interrogation of the underpinning racialized violence so central to Du Bois's work.[1] In repeating Du Bois's phrasing, Helmreich washes out the specificity of Du Bois's discussion and language. Race, it is implied here, is a twentieth-century problematic, receding beneath the rising tide of a "fresh problem" that "faces humanity now," in the midst of accelerating ocean acidification, the Earth's sixth mass extinction event, and ever-intensifying climate catastrophes that define the Anthropocene contemporary (Helmreich 2016, xxi). What we read in Helmreich's movement from Du Bois's color line to the twenty-first century's water line is a repetition of the critical lapse that has long plagued certain environmental humanities responses to the Anthropocene's exigencies, specifically, the gesture to a universalism that obfuscates the uneven distribution of risk and the differential impacts and experiences of environmental crisis.

We begin with Helmreich's borrowed phrase not to castigate or scold, but rather to call attention to the assumptions and presuppositions we bring to our work as cultural scholars engaging with the concept of *blue extinction*. What histories, epistemologies, and worldviews are privileged, suppressed, or rendered un/intelligible in the critical frameworks and archives that anchor our analysis? In his most recent monograph, for example, Helmreich makes a point to think with Caribbean and Black Studies scholars, naming Christina Sharpe's landmark concept of the wake and Barbadian poet Kamau Brathwaite's "tidalectics" as critical

[1] Du Bois is only mentioned one other time in the book, discussing race as "a group of contradictory forces, facts and tendencies" (qtd. in Helmreich 2016, 186).

engagements with watery spaces from which he "take[s] lessons" (Helmreich 2023, 9). Our goal in what follows is to amplify these somewhat submerged strains of blue humanities analysis by looking, specifically, to Black art that critically engages the oceanic as a foundational element of its aesthetic and material practice. In other words, how do we apprehend the twinned and inseparable violences enacted through both the water and the color line?

To think of the water and the color line as inextricable is to tune to an ambient current in Black Studies. At the same time that the ocean is the site of the Middle Passage's ongoing violence and the mass grave of Toni Morrison's "sixty million and more," Black artists in the diaspora have been sounding out a nascent oceanic ecology that calls into presence alternative figurations of Black futurities as ontologically promiscuous and capacious. Departing from the logics that underwrote a nearly Black and blue extinction, our chapter aims to model a critical methodology that, following Katherine McKittrick, thinks in a manner "disobedient not undisciplined" (McKittrick 2021, 232 n2) and that engages in the "generating and gathering of ideas—across-with-outside-within-against normative disciplines—that seek out liberation within our present system of knowledge" (54). In plumbing the critical currents that move around and through the concept of blue extinction, this chapter also takes a cue from Joshua Bennett, whose recent research attempts to "elucidate a divergent approach to the work of thinking at the intersections of black studies, animality studies, and ecocriticism" in order to articulate "a black hydropoetics that … makes hazy the division between person and nonperson so that a more robust, ethical lexicon for black life might rise to the air" (Bennett 2020, 173–174). In other words, in what follows, we are guided by the transits between life, water, and sound; it is our hope, however, that an emphasis on Black thought and Black sound might productively unsettle the intelligibility of these concepts.

Particularly in the context of articulating an ethics and ontology for the Anthropocene, we argue that critical engagements with oceanic spaces must grapple with the intimate entanglement of empire and colonialism with ways of mapping, representing, theorizing, and understanding the sea. From the outset, the oceanic turn in the humanities and social sciences has been wrapped up with the acquisitive thrust of empire. This is Elizabeth DeLoughrey's claim in her historicization of the emergence of the field of ocean studies. DeLoughrey traces the history of the "oceanic turn" to "Cold War geopolitics" and the 1945 Truman Proclamation, which kicked off a global scramble of territorial claims over ocean spaces (DeLoughrey 2017, 242). Following DeLoughrey's narration of the unfolding of the discipline allows us also to underline those lineages of blue humanities or ocean studies that foreground the place of diasporic and postcolonial thought to this field's emergence, including, in her example, Paul Gilroy's *The Black Atlantic* and the range of critical and cultural engagements with the ocean in Caribbean thought. In other words, instead of positioning technological advancements and recent scientific discoveries as a kind of moment of rupture that cracks open our calcified ways of imagining the possibilities or shapes for life, human/non-human relations, and kinship, might there be something to learn from thinking with the long ebb of existent oceanic thought? And how does the tenor of our techno-inflected vision of the world-to-come change when we remind ourselves that it emerges always from within a critical and theoretical framework defined by empire and colonialism, that is, marked by anti-Blackness and white supremacy?

Littoral Listening

To begin our process of disobedient gathering-generating-seeking, we turn first to Helmreich's *Alien Ocean* (2009), a kind of lodestar for a brand of Anthropocene-inflected critical ocean studies. This now-canonized text sets out to examine the ways recent transformations and discoveries in marine microbiology, particularly in relation to the presence and function of ocean microbes, have reconfigured longstanding cultural, social, and field-specific conversations around the possibilities for representing our relation to watery spaces and non-human others. For Helmreich, the alien is a concept, a "channel for exchange between the oceanic and the human, a transfer point between an alternately embracing and enigmatic ocean and us" (Helmreich 2009, xi). These alien microbes "are life forms whose place in our forms of life is yet to be determined"; they "mess up lineages and confuse the unfolding of chronologies" (Helmreich 2009, 17). The alien as transformative figure—as some *thing* that reorients our ways of being and relating to the ocean, to one another, and to the non-humans that live with and among and inside us—floats awkwardly alongside the recognition, a kind of bothersome reminder, that the ocean is also "a space of imperialism" and the site of "the Middle Passage" (Helmreich 2009, 12). Here, then, is the first point of pressure, an opportunity to amplify an undercurrent of Helmreich's conceptual apparatus. The oceanic alien embodies what DeLoughrey identifies as the "uncanny" contradiction of both an at-homeness and the absolute other (DeLoughrey 2017, 245); it "mess[es] up lineages" (Helmreich 2009, xi) and undoes chronologies. How might this figure be transformed if we understand it as always already related to, and emerging from, imperial cartographies and the world-shattering anti-Black violence of the Middle Passage?[2] In what follows, we argue that an attunement to something akin to a "blackened"[3] oceanic alien allows us to parse, or sound out, an *alien rhythm*, that is, a way of being in this world that is not quite of this world, but that nonetheless honors submerged histories and orients our thought toward liberation.

To begin to trace out this rhythm, we want to listen briefly to a perhaps unexpected tune, the Jamaican-born American writer and poet Claude McKay's 1929 novel, *Banjo: A Story Without a Plot*. The book's focus on interwar Marseille and its status as a littoral and liminal port city figures the ocean as a repository of extractive violence. Late in the novel, Ray, one of *Banjo's* two protagonists, looks out at the water and performs a close reading of the nascent racial capitalism emerging at sea:

> Ray's hankering was for scenes of tropical shores sifted through hectic years. Salty-warm blue bays where black boys dive down deep into the deep waters, where the ships shear in on foamy waves and black youths row out to them in canoes and black pilots bring them in to anchor. Coconut palms like sentinels on the sandy shore. (284)

This scene, and many throughout the picaresque novel, suggests a dual understanding of the ocean. On the one hand, there is a slight slippage into a mode of watery pastoral that aestheticizes and abstracts the moment

[2] Bennett (2020), in fact, already makes a similar claim about the "myriad forms of uncanny life" that dwell in the ocean's "capacious, irreducible blackness" (170). For Bennett, in terms that resonate with our efforts here in this chapter, that "uncanny life" opens onto a way of thinking about how Blackness has been instrumental for "organizing both human and non-human life (that is, the color line as the human-animal divide by another name)" (170). Our sense of the alien as a generative figure, we should note, contrasts with the dehistoricized figure of the alien that emerges in what Stacey Alaimo (2020) critiques as "settler-colonial visions of the ocean as an alien space" (317).

[3] We gesture here to Rinaldo Walcott's (2003) concept of "blackening," a term that describes the construction of Blackness through a complex push and pull between, on the one hand, violent and systemic imposition of racial logics and, on the other, tactical modes of resistance to that logic and self-expressions of Blackness (125).

of colonial contact.[4] Instead of slave traders and merchants docking, these are ships manned and moored by Black people in an egalitarian tropical economy. But, on the other hand, Ray knows that these Edenic shores are a fantasy—before we get to the diorama of coconut palms, we must pass through the marine violence evoked by those "hectic years."

Read in this way, we can see McKay rehearsing a mode of Helmreich's bluewashing in germinal form: there is a submersion beneath the waves of the problematic of race that only ever appears as a repressed force on the novel's metaphors. But McKay reveals this smokescreen for what it is elsewhere. Consider another passage from the novel:

> There any day he might meet with picturesque proletarians from far waters whose names were warm with romance: the Caribbean, the Gulf of Guinea, the Persian Gulf, the Bay of Bengal, the China Seas, the Indian Archipelago. And, oh, the earthy mingled smells of the docks! Grain from Canada, rice from India, rubber from the Congo, tea from China, brown sugar from Cuba … Barrels, bags, boxes, bearing from land to land the primitive garner of man's hands. Sweat-dripping bodies of black men naked under the equatorial sun, threading a caravan way through the time-old jungles, carrying loads steadied and unsupported on kink-thick heads hardened and trained to bear their burdens. (67)

This passage starts with McKay ironically invoking a similar romanticism, or even orientalism, to the section above—the water is aestheticized to evoke placid, salty dreams of exoticism. But just as quickly, we follow the jetsam to limn their imbrication within structures of Black exploitation. McKay's passage, in other words, ironically rehearses a colonial cultural imaginary that collapses Blackness into animality, transforming "black men" into beasts of burden. In its entanglement with the movement of commodities, the ocean creates a churning mass of actants and assemblages. But even as McKay's port appears as a utopian space of mixture and hybridity, the text refuses to perform an ontological flattening where the horizontal vibrancy of matter is celebrated without an acknowledgment of how matter always moves through a history of racialized violence. Toward the end of the novel, McKay drives this point home:

[4] Hannah Freed-Thall (2021) similarly notes McKay's interest in the "'picturesque' quality of the industrial port city" (142).

> The eternal harvest of the world on the docks. African hard wood, African rubber, African ivory, African skins. Asia's gifts of crisp fragrant leaves and the fabled old spices with grain and oil and iron. All floated through the oceans into this warm Western harbor where, waiting to be floated back again, were the Occident's gifts. Immense crates, barrels, cases of automobiles, pianos, player-pianos, furniture[.] ... Composite essence of the soil of all lands. (307)

The ocean becomes a palimpsest of the scriptures of a nascent petromodernity that depends upon racialized exploitation, hinging upon both ports and the "Occident's gifts." The anaphora of the second sentence emphasizes the slippage between thing and person, as well as the fact that the ocean itself attests to something like the fungibility of the Black body, unable to differentiate between "African hard wood" and "African skin." The ultimate suggestion here is that any aesthetic rendering of the ocean remains saturated with anti-Black violence. Black extinction is, in other words, the possibility that gives rise to any critical account of the blue humanities.

But what interests us in this chapter is not only a critique of the blue humanities through its lingering specter of Blackness, but how we might listen to the ocean otherwise to imagine a Black *and* blue humanities. *Banjo* models this critical modality as McKay interrupts the seamless circulation of capital, creating an insurgent inhabitation of the ocean by using it as a site to forge an alternative politic. Crucially, this is an acoustic fugitivity, stylized in the novel as "jazzing," what Michelle Stephens calls "a specifically black way of being in and expressing the condition of maroonage in the world of the early twentieth century" (Stephens 2005, 184). In *Banjo*, this takes place at two levels. There is the literal jazz that Banjo and his gang play, but jazz also appears as an epistemological proposition—Banjo's gang are continually looking for ways to sound out improvisatory riffs. They live on the margins and fringes of Marseille's maritime economy—they steal and beg for food off harbored ships, pilfer barrels of wine by putting straws into casks, sign up to join crews only to disappear with advance wages. Jazzing, in other words, points to the provisionality and experimentation of the novel's politics and form. Throughout the novel, there are a variety of political positions that are rehearsed and contested—from American expats toying with Garveyism to student revolutionaries flirting with fascism. As Anthony Reed puts it, "[*Banjo*] attempts to imagine new ways of being together and forms of politics beyond the limits suggested by the 'Negro intelligentsia' its characters criticize or those available under liberal capitalist modernity" (Reed 2013, 758). This is precisely the point. The suggestion here is that, yes, the ground of politics needs to be constituted in and through the ocean, but that there is also a polysemous possibility for what might emerge, a variety of contrapuntal rhythms threatening to erupt at any time.

To put this differently, McKay doesn't just show us how the ocean perpetuates colonial and anti-Black logics (which, of course, it does), but he creates a larger focus on how there might be different ways to listen to that ocean. This is true not just within the novel, but, as suggested by the subtitle, "A Story Without a Plot," in the novel's form. Formally, the novel is structured around a counterflow that seeps into the port through imported rhythms. As McKay's easy slippage between "African rubber" and "African skins" suggests, the commodities that travel across the ocean are defined by anti-Black violence. But alongside this violent circulation is Black music. We read, for example, that Banjo and his gang "played the 'beguin,' which was just a Martinique variant of the 'jelly-roll' or the Jamaican 'burru' or the Senegalese 'bombe'" (McKay 2018, 96). These rhythms interrupt the prose of the novel, pushing the narrative from the diegetic to the rhythmic. This reading of *Banjo*'s rhythmic form and its plotlessness anticipates Edouard Glissant's examination of "the problem of the rhythmic structure of the literary work" (Glissant 1999, 106). Glissant here considers the specificity of national literatures, as well as their emergence in the so-called postcolonial moment. Musing on the narratological structure of Western literature, with its "neutral zones of narrative" and periodic "explosive flashes," he suggests that, in a Caribbean context, "[t]he unvarying season (the absence of a seasonal rhythm) leads to a monotony, a plainsong whose obsessive rhythm creates a new economy of the expressive forms" (Glissant 1999, 106). Against the imposition of colonial form and Sameness—in the terms of our chapter to this point, a kind of universalism that obfuscates difference—for Glissant, the lived experience of the Caribbean multiplies the possibilities of alternative modes of textual representation.

Take the way the jazz number, "Shake That Jelly Roll," is integrated into McKay's novel. The narrative begins to crumble as the song slowly takes over the chapter: "Sing, Banjo! Play, Banjo! Here I is, Big Boss," we read as a crowd of "Seamen from Senegal. Soldiers from Madagascar. Pimps from Martinique" (McKay 2018, 49) converge onto the Café African bar where Banjo is playing:

> Shake to the loud music of life playing to the primeval round of life. Rough rhythm of darkly-carnal life. Strong surging flux of profound currents forced into shallow channels. Play that thing! One movement of the thousand movements of the eternal life-flow. Shake that thing! (McKay 2018, 57)

Here we see the port as a chokepoint, but instead of container ships coming in and out, McKay reimagines the port as the site where musical currents are "forced into shallow channels." We can see this in the modified lyrics to the song that are filtered through the oceanic: "Dry land will nevah be my land / Gimme a wet wide-open land for mine," Banjo sings (McKay 2018, 48). Even the tense changes in the chapter—moving from the past tense of the novel into a participatory presentism. McKay models a different way of reading the ocean by bringing to bear the musical energies of the Black diaspora that do not tune themselves to the rhythm of capital.[5]

What we are reaching for in this reading of McKay, to return to our earlier terms, is an alien rhythm that emerges in and through the ocean. Describing the violence of the Middle Passage, Fred Moten and Stefano Harney write of a killing rhythm that imposes itself on the world: "It asserts that nothing else can be heard, that nothing else need be felt" (Moten and Harney 2021, 55). At the same time, though, they also point to the way alternate acoustics emerge within this space: "But this rhythm has always been set amidst, and beset by, the general antagonism, the cacophony of beats, lines, falsettos, and growls, of hips, feet, hands, of bells, chimes, and chants, an undercommon track" (Moten and Harney 2021, 55). This is what we see emerge in McKay's formal introduction of Black diasporic rhythmic relations—a way of thinking through, with, and alongside the ocean that points to this cacophonous and riotous possibility. It is this same possibility that enables Glissant to trace the "cry of the Plantation" at the same time that he sounds out an alternate "rhythmic rhetoric of the shore" (Glissant 1997, 73, 122). We name this alternate and furtive rhythmic possibility the alien—a way of delineating a rhythm that is both in this world, but not of this world.

[5] Anthony Reed (2021) makes a similar argument, claiming, "the tom-tom and ideas of African rhythm stand in for the repaired temporality whose rupture is the Middle Passage" (35).

Alien Rhythm

Our turn to the alien in the context of blue extinction draws, of course, from the conceptual framework laid out by Helmreich and named above. But, more importantly, as we sound out the contours of this alien rhythm, the concept also circulates across Black sound and music criticism, often attached to artists and acts that have been understood under the umbrella of Afrofuturism. As one particularly rich example, we look to a 1992 piece of music criticism by Mark Sinker, published in *Wire* magazine and, for our purposes, serendipitously titled "Loving the Alien." Sinker's astonishing analysis draws connections across Black Science Fiction, Black music from New York hip hop and Detroit Techno to soul and gospel, and the tens of millions of individual histories of abduction and upheaval, death, and life in the wake, that are so intimately entangled with the ocean and the forced transoceanic movement of enslaved Africans. Sinker begins with the claim that has echoed across more recent critiques of the language of rupture or newness that attaches to some strains of Anthropocene discourse: "The central fact in Black Science Fiction ... is an acknowledgement that Apocalypse already happened: that (in PE's [Public Enemy's] phrase) Armageddon been in effect" (Sinker 1992, 31). Rescaled to Helmreich's microbes, we might restate this line of argumentation to claim the oceanic alien as a presence already known, understood, and experienced intimately; as a figure that has already circulated in Black American and Caribbean epistemologies and cultural artifacts, a figure that points back to that moment of contact, named in the epigraph above, of "the ships" that "landed long ago" and the "*Alien Nation*[s]" they augured and left in their wake. Sinker's analysis even identifies those contradictory, uncanny affects—the "affirmative, soul-gospel aspect" that articulates not only resilience but full-throated life in the face of "disaster," and the "dark, crazed, visionary tongues" that offer sideways or slantwise views of possible futures.

We close this chapter then by turning to one such "dark, crazed, visionary" Black sound project—the electronic music group Drexciya, which was composed of the late James Stinson and Gerald Donald and active in the Detroit techno scene between 1992 and 2002. Mick Harvey positions the group as "a sustained aesthetic and fictional performance" with "an indigestible core at the heart" of the project (Harvey 2020, 136). In terms that speak to our interest in the group's musical experimentation, Harvey is suggesting that Drexciya has, at once, resisted incorporation into the machine of pop sound and not been adequately engaged critically and theoretically. Indeed, much of the criticism that invokes Drexciya begins and ends with the mythology that the group articulates in the sleeve notes of their 1997 compilation album *The Quest*. Those notes, attributed to a figure named "The Unknown Writer," make clear the speculative myths of the duo's origin: Drexciya is an underwater country peopled by Drexciyans, aquatic post-human beings descended from enslaved, pregnant Africans thrown overboard during the Middle Passage: "During the greatest Holocaust the world has ever known, pregnant America-bound African slaves were thrown overboard by the thousands during labor for being sick and disruptive cargo. Is it possible that they could have given birth at sea to babies that never needed air?" (qtd. in McKittrick 2021, 60). The notes end with a series of questions: "Do they walk among us? Are they more advanced than us? How and why do they make their strange music? What is their quest?" (qtd. in McKittrick 2021, 60).

The Drexciyan mythology is an evocative and theoretically rich conceptual framework for an experimental sonic act, and critics have dutifully examined its resonances. Kodwo Eshun, an early champion of the duo and the first to draw a connection between their mythology and Gilroy's *The Black Atlantic*, for example, describes the project as "inventing another outcome for the Middle Passage" (Eshun 1998a). More recently, in a study of a range of Afrofuturist methodologies, Robyn Maynard argues that Drexciya "intervenes in the history of the black Atlantic in order to reimagine a more generative narrative of the Middle Passage" (Maynard 2018, 36). We follow Eshun, Maynard, and others in our understanding of the group's self-articulated mythology as an attempt, from the very deep of nearly unfathomable suffering and violence, to identify plenitude, a range of Black futurities, and a mode of Black being that not only persists but becomes something altogether different, outside of the scripts imposed on it by whiteness, empire, and colonialism. In our estimation, this is the key feature of the Drexciyan rereading or reimagining of the Middle Passage: in the creation of Drexciya and the mutations of Drexciyans, Blackness thinks and constructs itself on its own terms, that is, not as something emerging from the theft, displacement, and violence of slavery, but as an entirely new articulation of an almost-here that disorganizes the efforts of colonial, epistemological frameworks to define, contain, and subdue.

This gloss on Drexciya's mythology shifts our understanding of the creative work of the duo from flatly resistant and oppositional to something more, something active, and resonant with McKittrick's broader argument about the need to recognize that, while Black music often performs "lyrical and sonic critiques of colonialism, racism, structural inequalities, and other forms of violence … something else is going on that cannot be easily captured by an analytic of oppression/resistance" (McKittrick 2021, 57). Indeed, McKittrick's reading of Drexciya as not only about the movement from the Middle Passage toward an aquatic, Afrofuturist utopia-to-come, but rather as "collaborative sound-labor that draws attention to creative acts that disrupt disciplined ways of knowing" has been transformative for our understanding of the group's mythology, music, and its method of production (McKittrick 2021, 61). This is a vital intervention into the extractive analogies of Helmreich we glossed earlier, a way of thinking through how Drexciya performs an acoustic intervention that unsettles intelligibility—of genre, of discipline, of musical or sonic content, and also of the racializing scripts that underwrote the violence of the Middle Passage and that continue to circulate as white supremacy in the present moment. This is partly an argument about the intelligibility of the body behind the machine, that is, what Eshun identifies as the technological capacity of "sequencers, samplers, synthesizers, and software applications [. . .] to scramble the ability to assign identity and thereby racialize music" (Eshun 2003, 296).[6] The fundamental opacity at the core of Drexciya—the musicians involved were not identified until after Stinson's death—becomes apprehensible as a practice of fugitivity, an attempt to keep open a range of identities and futures to move through and across, a way, echoing Eshun's broader argument about the "cyborg fantasies of Detroit techno producers … both to alienate themselves from sonic identity and to feel at home in alienation" (Eshun 2003, 296).

Beyond remaining mired in the Drexciyan mythology, another pitfall of criticism that engages the group is a reticence to actually *listen* to the music.[7] And so, we conclude with an attempt at a kind of close listening

[6]This relates back to Sinker who similarly suggests that "Techno, Detroit's 80/90s black electrowave, explicitly and contemptuously refused community with Motown and motorcity gospel for Gary 'Me I Disconnect From You' Numan" (Sinker 1992, 32).

[7]Many critics, too, perform wonderfully evocative readings and textual renditions of Drexciya's specific brand of techno which have shaped how we listen to, and hear, their sound. Cf. Eshun, across all of his writing about the duo; McKittrick 2021, 61–62; Harvey, who describes "a remarkably coherent gestalt" in "the Drexciyan sound," going so far as to identify the presence of "oscillator cross modulation," that is, "an obscure feature of a handful of Japanese synthesizers produced in the 1980s" that, in his estimation, "spans virtually all of [Drexciya's] diverse output" (Harvey 2020, 138); Williams 2001, 168; and Thomas 2022, 53.

with particular emphasis on the question of intelligibility and the track "You Don't Know" from the 1996 EP *The Return of Drexciya*. Even the title of the EP destabilizes sense. "Return" here is both the reappearance of the group, and might be paraphrased as "another release by Drexciya," but also the Drexciyan movement of return to Africa that the group states clearly in the sleeve notes to *The Quest*.[8] Put differently, the EP's title makes plain the plenitude of meaning that, in our analysis, runs through all of the Drexciyan catalogue, from the music itself and the broader mythology to the paratexts of the liner notes and scattered interviews. A simple phrase—the return of Drexciya—carries within it the echo of that slantwise, visionary reading, understood in the "will have been" of the future perfect tense as not only anticipating, but calling into presence a futurity that is also liberation.

"You Don't Know" begins like many Drexciya tracks: a dark, rubbery, driving 808 drum machine forges ahead under a jittery hi-hat holding time. After eight bars of this rhythmic repetition, a new sound bubbles up from under the track, something like the slow and intentional inhales and exhales of breathing underwater with a scuba apparatus. Another eight bars and the song is interrupted by a sequence of bright tones that shimmer across the rhythm, evoking the fast twitch movements of underwater creatures or sunlight filtering through to the ocean's depths, distorted by surface waves. The song's primary melody enters next, an arpeggio of five notes like a pointed question, or perhaps verging on a kind of alarm. And then, like a whisper underneath the music, the song's single lyric appears barely audible and begins to repeat intermittently: "You don't know what the fuck I'm talking about." From there the song drives forward to its close, remixing its melodic and sonic elements before introducing new effects in the final third like whooshes of compressed air or the drag of water on a submarine body. "You Don't Know" recirculates many of Drexciya's signature sounds and sonic affects, from its "squelchy atmosphere" and "sequencer patterns that sound like bubbles rising to the surface of the ocean" (Williams 2001, 168) to its "acrid frequencies," "abrasive tones," and "jagged snare velocities" (Eshun 1998b, 83).

[8] The liner notes from Drexciya's 1997 compilation album *The Quest* show the world map in four stages, describing the displacement and forced migrations of Black people, the global dissemination of Black techno, and the future "journey home" to Africa. Cf. Williams 2001, 169.

Drexciya's "You Don't Know" crystallizes the disciplinary unsettling to which we hope our chapter productively, and respectfully, contributes. "You don't know what the fuck I'm talking about" is a phrase confrontational and oppositional—it marks a difference and constructs an interior space from which the addressee is excluded. The phrase embodies that "lyrical and sonic critique" that McKittrick identifies as a foundational element of Black musical texts. But it is also a phrase that might be read further, as an assertion of the Drexciyan project of unsettling intelligibility and disorganizing disciplinary thought in and through techno music and a critical and speculative engagement with the subaquatic histories of the Black Atlantic. It is, explicitly, not that you don't know what the fuck *you*'re talking about. Instead, the phrase announces the coming into existence of an epistemology that is *alien* to the Drexciyan speaker's interlocutor. This is an epistemology of this world and not of this world, that draws from ancestors and appears like something altogether new—a seeming contradiction that might be illuminated by Drexciya's method of production. As McKittrick tells us, the duo "played live into a predigital analogue recorder. What we are given, as listeners, is synthesized improvisation" (McKittrick 2021, 62). Here, then, is another kind of "return," to the jazz improvisation of *Banjo* and McKay's port city protagonists. An improvisatory musical practice becomes a method for living, a provisional and fugitive stance on the world that always keeps open the possibility of movement, of escape, of return to an almost-legible world-to-come.

References

Alaimo, Stacey. 2020. Afterword: Adequate Imaginaries for Anthropocene Seas. In *Blue Legalities: the Life and Law of the Sea*, ed. Irus Braverman and Elizabeth Johnson, 312–325. Durham and London: Duke University Press.

Bennett, Joshua. 2020. *Being Property Once Myself: Blackness and the End of Man*. Cambridge and London: Harvard University Press.

DeLoughrey, Elizabeth. 2017. The Oceanic Turn: Submarine Futures of the Anthropocene. In *Humanities for the Environment: Integrating Knowledge, Forging New Constellations of Practice*, ed. Joni Adamson and Michael Davis, 242–258. London and New York: Routledge.

Eshun, Kodwo. 1998a. Drexciya: Fear of a Wet Planet. *The Wire*, 167. https://www.are.na/block/1504716. Accessed 4 April 2024.

———. 1998b. *More Brilliant than the Sun: Adventures in Sonic Fiction*. London: Quartet Books.

———. 2003. Further Considerations on Afrofuturism. *CR: The New Centennial Review* 3 (2): 287–302.

Freed-Thall, Hannah. 2021. Beaches and Ports. *Comparative Literature* 73 (2): 131–150.

Glissant, Edouard. 1997. *Poetics of Relation*. Translated by Betsy Wing. Ann Arbor: The University of Michigan Press.

———. 1999. *Caribbean Discourse: Selected Essays*. Translated by J. Michael Dash. Charlottesville: University Press of Virginia.

Harvey, Mick. 2020. *Harnessed the Storm*—Rereading Drexciya with *The Black Atlantic. Studies in Gender and Sexuality* 21 (2): 136–140.

Helmreich, Stefan. 2009. *Alien Ocean: Anthropological Voyages in Microbial Seas*. Berkeley, LA and London: University of California Press.

———. 2016. *Sounding the Limits of Life: Essays in the Anthropology of Biology and Beyond*. Princeton and Oxford: Princeton University Press.

———. 2023. *A Book of Waves*. Durham and London: Duke University Press.

Maynard, Robyn. 2018. Reading Black Resistance through Afrofuturism: Notes on Post-Apocalyptic Blackness and Black Rebel Cyborgs. *Topia* 39: 29–47.

McKay, Claude. 2018. *Banjo: A Story Without a Plot*. London: Forgotten Books.

McKittrick, Katherine. 2021. *Dear Science and Other Stories*. Durham and London: Duke University Press.

Moten, Fred, and Stefan Harney. 2021. *All Incomplete*. Colchester, NY and Port Watson: Minor Compositions.

Reed, Anthony. 2013. "A Woman is a Conjunction": The End of Improvisation in Claude McKay's *Banjo: A Story Without a Plot. Callaloo* 36 (3): 758–772.

———. 2021. *Soundworks: Race, Sound, and Poetry in Production*. Durham and London: Duke University Press.

Sinker, Mark. 1992. Loving the Alien. *The Wire* 96: 30–33.

Stephens, Michelle. 2005. *Black Empire: The Masculine Global Imaginary of Caribbean Intellectuals in the United States, 1914–1962*. Durham and London: Duke University Press.

Thomas, Tashima. 2022. An Ecocritical Look at Flint's Water Crisis and Afro-Gothic Liquidity. *liquid blackness* 6 (2): 38–59.

Walcott, Rinaldo. 2003. *Black Like Who?: Writing Black Canada*, 2nd rev. ed. Toronto: Insomniac Press.

Williams, Ben. 2001. Black Secret Technology: Detroit Techno and the Information Age. In *Technicolor: Race, Technology, and Everyday Life*, ed. Alondra Nelson and Thuy Linh N. Tu with Alicia Headlan Hines, 154–176. New York and London: New York University Press.

Index

© The Author(s), under exclusive license to Springer Nature Switzerland AG 2024

V. Fibisan, R. Murray (eds.), *Blue Extinction in Literature, Art, and Culture*, Palgrave Studies in Animals and Literature,
https://doi.org/10.1007/978-3-031-69910-8

SPRINGER NATURE

GPSR Compliance

The European Union's (EU) General Product Safety Regulation (GPSR) is a set of rules that requires consumer products to be safe and our obligations to ensure this.

If you have any concerns about our products, you can contact us on ProductSafety@springernature.com

In case Publisher is established outside the EU, the EU authorized representative is:

Springer Nature Customer Service Center GmbH
Europaplatz 3
69115 Heidelberg, Germany

The manufacturer's authorised representative in the EU is Springer Nature Customer Service Centre GmbH, Europaplatz 3, 69115 Heidelberg, Germany. If you have any concerns regarding our products, please contact ProductSafety@springernature.com

Printed and bound by CPI Group (UK) Ltd, Croydon, CR0 4YY
30/09/2026
02225502-0004